The Book Cover

Spherical growth of the universe continues as corporeal matter is precipitated behind the outward progressing deflagration wave. The wave continuously penetrates farther into primordial space and matter. We can observe only half the distance to the newest corporeal matter as it is initially advancing outward at the same speed as light from that new corporeal matter radiates in our direction. Matter, Primordial and Corporeal, are of equal density. (Consistent with First Law of Thermodynamics)

We may be as close as three and one half Bly from the initial annihilation, maybe much farther. (Topics 1 and 4; "Time ..." and "Deflagration ...")

MODEL OF THE UNIVERSE

This "Model of the Universe" combined with the "New Universe Theory" [Ref 1]; give a complete explanation of the origin of all physical matter. This includes all mass and energy. It does not explain spiritual, religious, or biological origins

by

Bobby McGehee

authorHOUSE®

AuthorHouse™
1663 Liberty Drive
Bloomington, IN 47403
www.authorhouse.com
Phone: 1-800-839-8640

First published by AuthorHouse 2/10/2010

ISBN: 978-1-4490-6794-6 (e)
ISBN: 978-1-4490-6792-2 (sc)
ISBN: 978-1-4490-6793-9 (hc)

Library of Congress Control Number: 2009914336

Printed in the United States of America
Bloomington, Indiana

This book is printed on acid-free paper.

Synopsis

This scientific **Model of the Universe** includes primordial matter and the space surrounding the universe. It explains the universe's origin and continuing growth. The term "Universe" comprises everything that physically exists: the entirety of space and time, all forms of matter, mass and energy, including momentum, and the physical laws and constants that govern. This <u>Model</u> of the Universe uses only concepts and processes that are consistent with <u>Proven Facts</u> and the <u>Laws of Physics</u>. Every building block is supported by documented observations by credible astronomers, astrophysicists, and engineers. (See 'references' and 'credits') This Model is not based on any presumed, unproved phenomena, or hypothetical concepts.

The Universe Model explanations are intended to be totally understandable by all. The appendix includes a glossary for understanding terms that are used. Most equations and math explanations are not included in the text, but are relegated to the appendices for those that wish to pursue the topics and the proof.

The universe is enormous, and humongous; it is so large it is difficult to conceive in our mind. Thanks to the 1912 red shift* discovery by V M Slipher at Lowell Observatory, we have a method of measuring the age, growth, and distances to far away objects. There is much more to learn.

The book "New Universe Theory with the Laws of Physics [1]" is a frequent reference because it provides analogies for concept understanding.

It is highly recommended the Glossary and the Pertinent Laws of Physics for Origin of the Universe be reviewed before reading the Model text. (Appendix 2 & Appendix 6) It is physically impossible for any process to occur inconsistent with the proven and never disproved Laws of Physics.

Free Preview

This preview provides busy, interested thinkers enough insight to justify personal time for considering this Model.

We deserve a scientific explanation for origin of the universe.

Since mankind looked into the nighttime sky, we have tried to answer the question, "Where did it all come from?"... In 1910-1914, Lowell Observatory's V M Slipher put a spectroscope on the telescope to study elements in stars. He discovered red shift and farther away objects are receding faster than closer ones. The appearance was as if everything originated from a single point 14.7 billion years ago and everything is accelerating outward. For the first half of the century, through about 1960, scenario composers could only base their story on observation. The Big Bang Theory was developed and accepted; and then construed to try to make it fit. However, since the mid 1900's discoveries and knowledge have been acquired that define our history. Robert Jastrow's Star analyses and Harvey Richer's globular cluster research team found faded white dwarf stars over 14 billion years old. Some of those had descended from 8 billion year old stars, whose ancestors were 2nd and 3rd generation stars. This credible evidence proves the universe is at least 28 billion years old.

True scientific thinking uses logic, with only proven facts and processes which are the Laws of Physics. Observations are reexamined and the answer that is consistent with 'Laws of Physics' is this "Model of the Universe". It is logical and supported by verified observations.

Bobby McGehee explains; This Model of the Universe consists of provable, facts and logic using only verifiable observations and knowledge. It includes accepted and proven:
Isaac Newton's 1620 Gravity; **Albert Einstein's** 1909 Relativity; **V M Slipher's** 1912 discovery that the farther away the next galaxy, the

faster it is receding; and **Paul Dirac's** 1927 discovery of Positrons and Positroniums.

McGehee, among others, never accepted the BB theory because the BB is based on unprovable ideas that are inconsistent with well established Laws of Physics.

"As a thinking person, do you think it is more logical to assume that everything in the universe came from a single point, than it is to think that all space was full of something that converted to everything that now exists? All a person needs to do is to recognize the possibility that the universe and its objects are <u>not accelerating</u> outward."

Figure 1

LAWS of PHYSICS Pertinent to Origin of the Universe

Apply at all times and places without limit. Cannot be violated.

All mankind's knowledge and understandings are based on Scientific Thinking.

Scientific Thinking Uses logic only with proven facts & verifiable phenomena
Laws of Physics Principles verified many times, but never disproved
Myths Cannot be proven or disproved. (Based on fantasy or wishful thinking)
Belief & Faith ...Accepting without proof. (e.g., BB)
Presumption ...Arrogant acceptance of belief based on assumption or supposition

Proven, and never disproved:

(1)Motion (Isaac Newton, 1610)
$F = ma, = d(mv)/dt$
Force required for acceleration

(2)Energy & Mass (Albert Einstein)
$E = mc^2$
Energy and mass are convertible

(3)Force of Gravity (Isaac Newton)
$F_g = m_1 \times m_2/d^2$
Mutual attraction all objects

(4)Relativity (Albert Einstein, 1905)
$m = m_0 / (1 - m^2/c^2)^{1/2}$

(5)Continuity (Rudolf Clausius, 1850)
1st law of thermodynamics
$M_1 = M_2$, mass & energy into a
system = mass & energy out + mass
& energy remaining in the system.
*Succinctly; Matter cannot be
created or destroyed. (No BB!)*

(6)Schwarzschild radius (1610)
$R_S = 2Gm / c^2$
Radius where escape velocity = < c

(7)Escape Velocity
$V_{esc} = (2Gm / r)^{1/2}$,
For a black hole: *$V_{esc} = c$
(*Impossible to achieve)

Nomenclature:

F ...	force
E ...	energy
m ...	mass
v ...	velocity (speed)
V ...	velocity (vector)
a ...	acceleration (~dv/dt)
c ...	velocity of light (light speed)
M ...	matter (both mass & energy)
e ...	entropy
r ...	radius
G ...	universal force of gravity $= 6.67 \times 10^{-11} \dots Nm^2/kg^2$
N ...	Newton unit of force (Kg meters / sec^2)
R_S ...	Schwarzschild radius

Preface:

This "Model of the Universe with Laws of Physics" concept is presented with comparisons to the previous Big Bang idea. Recent findings (since June 2009) are presented that support this continuously growing new model, and illustrate how misinterpretation of the 'Hubble Number' led to the 1920's postulation of a single-point source.

The BB theory is thought to be a spherical region that contains all matter, which erupted out of nothing from a single dimensionless point at a single infinitesimal instant into sufficient energy to produce all matter (energy and mass). The initial BB energy condensed into particles that coalesced into all of the galaxies, stars, planets, objects, and things, and is said to be accelerating / expanding in all directions. The BB model is based more on presumptions than proven facts.

The New Model View is that the universe is a spherical region within a much larger region of primordial matter. Primordial matter is made up of matter (+) and antimatter (-) particles stabilized by a hexahedron crystalline like arrangement. When positron + and electron − particles come in contact, they annihilate into photons; and as photon concentrations become adequate, it precipitates into corporeal matter of the universe. The initial annihilation started at a single + & - pair which upset positronium rotation synchronization. Photons from annihilations propagate in all directions and produce more chain reaction annihilations. Outward flowing Photon density increases to concentrate, coalesce, and precipitate into subatomic particles. Particle accretions produce the objects and matter of the universe. The objects decelerate outward behind the continuing process called the deflagration wave, (because a readily understandable analogy relates to a flame propagating through dry grass: matter is not created nor destroyed; it is only converted from one form of matter into another). All processes in this **New Model** have been proved, and verified; all are consistent with the **Laws of Physics**.

Introduction

Hubble confirmed Slipher's finding that distant galaxies are receding at higher velocities than closer ones. Also, this LOP Model proves velocities of all galaxies are decreasing with time. Thus the universe is growing, yet space is <u>not</u> expanding.

Since the beginning of consciousness, mankind has tried to explain the physical observations of the night sky. The late Carl Sagan found by his research that one early idea for the origin came from Scandinavians: "a giant and a cow laid an egg which hatched and …there was the universe!" Hypatia, Copernicus, and Galileo were more scientific and proved that the earth is not at the center and not flat, as some thought.

In the 1910's V M Slipher discovered that galaxy recession velocities increase with distance. In the 1920's and 30's it was erroneously hypothesized that galaxies were accelerating apart and the universe is expanding. It was then graphically shown by backward extrapolation everything appeared to come from a single point, which graphically appeared to be 13.7+ billion years ago. Sir Fred Hoyle called the explanation a "Big Bang."

Facts such as the known ages of progenitor stars are shown in the milestone bar chart, but time required for progenitor stars and Laws of Physics were disregarded by BB promoters. Their presumptions and assumptions led to the "Big Bang Theory." Unfortunately, the (BB) has been accepted and rationalized by several decades of astronomers.

The Model of the Universe explanation as presented herein was initially called "New Universe Theory," however, after further study and investigation it is now referred to as the Laws of Physics "<u>Model of the Universe</u>" because it is beyond theory and has proved to be valid. It is scientific and plausible because it uses only proven facts and is consistent with observations.
All observations better support this new concept than a BB.

Note:

This publication is called a supplement as it includes material 'intended' to be incorporated in the original NUT[1] writing, and it also includes observations and analyses that have been developed and acquired more recently. The original NUT concept went to press early to allow presenting it to several notable and credible aging Scientists who never accepted the BB theory for scientific reasons. Some BB believers were cruel and continue to speak disparagingly of those who refused to accept their BB idea, but forgive them. "They knew not what they did not know"; a Laws of Physics alternative had not yet been developed.

Internationally known life long BB rejecters; namely, Grote Reber, Fred Hoyle, Thomas Gold, and Hermann Bondi, had all deceased by June 2005, except for Hermann Bondi, who was first chairman of the ESO (European Space Organization). In an attempt to get this concept to Bondi while he was still cognizant, the first published copy of "New Universe Theory" was sent on September 1, 2006. However, it was returned 6 weeks later with a heartfelt note stating "undeliverable because addressee had deceased on 10 September."

A research idea (presented in angular momentum section / Abell ACO clusters) has been started and follow-on studies are proposed, the results of which will positively prove the validity of this New Model. It proves accelerating expansion and inflation do not exist; but growth does. (Therefore no BB)

Proof is included that demonstrates the universe is about 30 billion years old, over 60 billion light years in diameter, and is growing. The region of our residence in the universe is only a few billion light years away from the universe's start site.

Future astronomers are going to have lots of fun; researching and producing further verifications with their discoveries.

New Universe Model Topics:

Contents

<u>Why we need this credible New Model !</u>

1. Matter creation by the BB model violates the 1st law of thermodynamics: "Matter (mass and energy) cannot be created or destroyed."

2. Recession of galaxies with acceleration violates Newton's Laws of Motion: "F = ma"; there is no force. ('Dark energy' is strictly a presumption.)
Expansion is an illusion from misinterpretation and assumption that Red-Shift increases not only with distance but also with time.

3. Schwarzschild Radius (R_s) cannot be denied, or violated, by any form of matter; energy or mass: "Velocity cannot exceed the speed of light."
"Gravity is relentless." (e.g., Earth's R_s is < .9 cm)

4. The BB provides no explanation for origin of the huge amount of angular momentum that exists throughout the universe: Angular momentum originates from transferred linear momentum. The BB does not even consider momentum conservation!

5. Observation of faded white dwarfs in the M-4 Globular proves the universe is older than possible with the BB: (> 24+ billion years; BB limit is 13.7)

6. Abell (ACO) Galaxy Cluster Count Asymmetry by both time and direction proves the universe is growing and galaxies are decelerating in the direction away from the older region of larger clusters.

1. Milestone Chart: Universe Minimum Age

Understandable History of Time:

… Illustrated in chronological order; Ten time increments were necessary for the existence of the objects in our region of the known universe. *(Increments are itemized following the chart).*

Milestones above the main time bar identify chronological events since the first primordial matter annihilation. The sum of the times between milestones culminates into the time necessary to produce white dwarf faded descendants of third generation stars. Such stars are telescope observed in the M4 Globular Cluster.

The **FIRST TIME INCREMENT** was from '1st positronium annihilation' until the wave of annihilations reached our region.

The **SECOND INCREMENT** is time period while the ~3 Billion light year wide wave transcended our location. Within the wave, transformations from photons into elementary mass particles occurred. Under the force of velocity enhanced mutual gravity between adjacent particles, they fused progressively into larger particles. These new particles further fused into nuclides of protons, and neutrons (not by collisions from high pressure and temperature as presumed in the BB theory). Particles ultimately accrete into astronomical objects of all sizes. Particles never collide precisely Center of Gravity, to Center of Gravity; therefore some linear momentum is always converted to angular momentum.

The **THIRD INCREMENT** begins with the birth of the first star objects in our region. Many sizes of first generation stars appeared and most of these accumulated into galaxies, and galactic clusters, as found throughout the universe today. Mutual gravitational orbiting occurs frequently which consumes outward flow linear momentum. Angular momentum as observed throughout the universe substantiates this concept.

The **<u>NEXT SEVEN TIME INCREMENTS</u>** are the times consumed for star evolution to culminate into the faded white dwarf stars [24] observed in the M4 globular cluster <u>TODAY</u>.

Figure #3

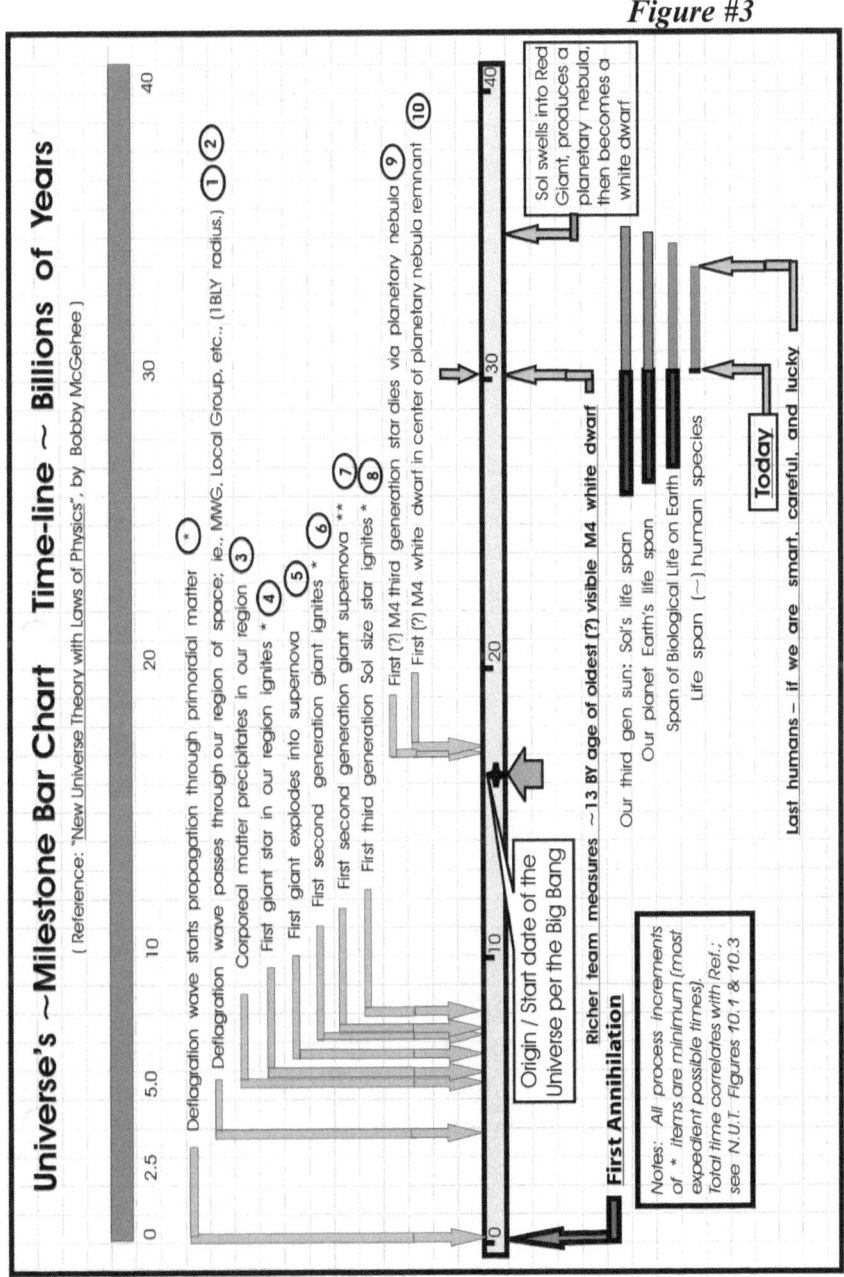

The lower side of the main time bar (Figure 3) illustrates the History and Future of our Solar System and Milky Way Galaxy. ... 'Today' universe age is ~ 30 Billion years.

Our solar system and galaxy are younger than the globular cluster M-4 white dwarfs, yet the galaxy constituent stars evolved into existence in the same manner as many, if not all, M-4 white dwarf progenitor stars. *(How the ages of these star systems are different is logically explained in the reference NUT [1]).* Our younger sun has only lived about ½ of its 8 to 10 billion year fusion life. The four bars below the overall universe bar illustrates the total life time for our sun, our earth, life on earth, and human existence. The last bar shows the maximum potential life span for our human species. The 'fat' arrow below and to the left of the center of the universe's time bar indicates when the BB proponents think the origin/start date of the universe began!

On a spatial distance scale we always will be slowly decelerating outward from the universe origin. Our physical location (radial direction and distance) in the universe is currently estimated (see Voids, 2007 discovery) to be between 3.5 and 7.5 Billion light years. Our galaxy will continue to circulate in and with the 'local group', vortexing in the 'three dimensional sea of space'. Our outward velocity from the universe's center will continue ad infinitum to decrease as vorticity and entropy increases to consume energy from the outward linear momentum. In the process our local group will merge with other galactic clusters ([1] Figure 10.3).

On the time scale, we are about 28 to 32 billion years from the universe's origin and start point, but our spatial location is much less. Recent observations (2000-2008) indicate there may be about 30 galaxies revolving and circulating in our local group. Our Local Group of galaxies is now marginally qualified to be listed as an ACO (Abell) cluster. (See Section 3 for more proof of asymmetry)

Ten Chronological Time Increments on the chart:

1. The deflagration wave reached our region and continues beyond. **(~ 3.5 By) (Billion years)**

2. Within the wave thickness primordial positronium matter is continuously converting into mass particles and objects as the wave travels onward and outward. Mass particles and objects precipitated into our region of the universe as it passed through. **(1.5 By)**

3. Continuing velocity enhanced coalescence and clumping of precipitated mass objects produced stars, including the most senior ancestral giant progenitor stars of M-4 White Dwarf stars. **(0.3 By)**

4. Fusion burning of fuel within these most senior giant stars was rapid and continued until they exploded in supernovae which disbursed material throughout the region. **(~.8 By)**

5. Some time is required for the supernovae debris to gravitationally collect into star masses. **(~.8 By)**

6. Debris collected into second generation rapid burning stars fused some of the light weight nuclides into heavier nuclides limited to the weight of iron until second generation star supernovae again dispersed that material by supernovae explosions. Heavier than iron element nuclides were formed by the extreme pressures and temperatures within and during the brief supernova explosions. **(.25 By)**

7. Some of the second generation giant ancestral progenitor supernovae debris collects and accretes into third generation stars, some with masses that are similar to our sun, between .9 and 1.44 solar masses. **(~ .6 By)**

8. Third generation stars ignite and their fusion processes continue for 8 to 10 billion years, like in our sun. Near the end of the fusion life of these third generation stars, they swell in size as fuel reserves are significantly reduced, resulting in less gravity to maintain compactness. They swell and cool as they become Red Giants. **(~ 8 to 10 By)**

9. At the end of their Red Giant phase, fuel reserves are depleted and these stars pulsate between nuclear fusion cycles of 'off-on-off'. Mass debris is expelled at each cycle of above and below critical mass pressure/density stage. Material is thrown off at each cycle producing planetary nebula (rings of glowing debris) while ultimately leaving at the core a dormant white hot 'White Dwarf' star. **(.01 By)**

10. White dwarf stars are totally dormant, no longer having sufficient mass and gravity pressure to support fusion, so they cool slowly, only by thermal radiation. Cooling to non-incandescence requires about 14 billion years. **(14 By)** [24]

Total time = 28 to 32 By

Conclusions from the "Time Line Bar Chart":

The milestone bar chart reveals our thread through time from the beginning of the universe (first annihilation) to, through and beyond the demise of the last living / biological things on earth. Why globular clusters are older than stars in the Milky Way host galaxy and yet populate the same region of the universe is no longer a mystery. It is plausible, logical, and is consistent with this origin of the universe concept (as explained in Reference [1]).

Several studies of 'faded' white dwarfs in star clusters are continuing. Two of these studies were started early in the last decade of the 20th century by two teams of astronomers, both headed by Harvey Richer of the U of B C, by observing the dimmest white dwarfs in the M-4 globular cluster [24]. The milestone identifying this observation is on the lower side of the time bar, and is labeled 'Richer Team'; also labeled 'Today'.

The only known way the human species can survive beyond the human species life time bar is if a younger planetary system can be found with a habitable planet. If we're lucky, we <u>think</u> we have ... maybe 4 ½ billion years before our time expires.

Distribution of chemical elements

Time / Age
Heavy elements (nuclides heavier than iron) were formed early in the universe, when the earliest second generation giant stars burst into supernovae; About 6 to 7 Billion years after the first annihilation starting event. (*Figure 3 is a time chart, not a distance map.*) The heavy nuclides were fused by the high pressure and temperature within the blast, which also dispersed them. That occurred within about one Billion years before the first solar size (.9 to 1.44 Sol) stars were precipitated from that debris, which was maybe within as little time as one billion years, which was ~ 8 Billion years after the initial annihilation of positrons and electrons.

Distance / Location
For directional and distance considerations, metals such as nickel, lead, silver, gold, platinum, as well as all other nuclides can be found everywhere between the origin, and all the way out to the newest second generation giant star supernovae which are within about 6 or 7 billion light years of the outward progressing deflagration front. The deflagration front is currently about 28 or more billion light years away from its origin site. However, we can only see stars precipitated at about ½ that distance because the front is moving away at almost the speed of light, and the light radiation from stars that are more than almost half way has not yet reached our region of the universe. (*Illustrated in Figure 8, and indicated on the book cover, both indicate physical distance relationships*). Due to the speed of the outward progressing wave / front and the speed of light travel, the farthest away we can observe physical objects and processes will appear to us as if they were recent.

2. Acceleration vs Deceleration; Asymmetry

Hubble numbers are the ratios of the radial component of velocity vs distance. Increasing red-shift with larger distances does not define or indicate acceleration! Red-shift is an approximate measurement of the radial component of velocity from us, but it does not, and can not support 'presumption of acceleration'. Acceleration or deceleration must be proved and determined by other means, including scientific logic. **The BB theory presumption** is that galaxies are accelerating along a Hubble line, yet there are **no forces for acceleration**! The NUT concept used observations, logic and scientific thinking to substantiate that a Hubble line is a gradient line across deceleration lines of different galaxies, not a line of galaxy travel. **Galaxies are in fact clustering and decelerating**!

Variation of Hubble Numbers as observed by astronomers in the 1900's was assumed by the author of the NUT to be due only to directional asymmetry. A table, [1] Figure 3.1 is a list of a range of Hubble numbers, all of which were measured by credible astronomers. A Hubble Number is defined as the numerical value of the ratio of a galaxy's recessional velocity, determined from Doppler red-shift measurement, divided by the distance to the galaxy as determined using various distance markers. When a line is drawn on a graph of velocity verses distance from zero distance and zero velocity through this number on the graph, it suggests the galaxy is traveling along a straight line 'with acceleration', in an outward direction along the 'Hubble line'. Acceleration is only a presumption. This is a process of major importance where the NUT concept is consistent with laws of physics and the BB is not.

Simple analogies were used (Reference [1]) to illustrate how the observation of increasing 'red-shift with distance' is simply deceleration with growth, not acceleration and /or expansion. One analogy describes a truck load of potatoes that are being spilled as the truck progresses along a road past an observer.

Individual spilled potatoes are observed and known to <u>decelerate while those closer to the truck have higher velocities at greater distances from the stationary observer</u>. Also, if a bug on any one of the potatoes observes potatoes in <u>both</u> directions, they all appear to be accelerating away from it. However, the potatoes are undeniably separating … yet <u>decelerating</u> apart, just like galaxies of the universe are decelerating from us and apart from each other, while being <u>mis</u>interpreted as accelerating.

It has been suggested that the rate of separation in forward and reverse directions from any reference potato might be detectably different (revealing asymmetry). Measuring Hubble number directional asymmetry might allow us to determine our location in the universe. The asymmetry assumption might be verified and calibrated if deceleration is non-linear with distance, as it is in all other fluid flows. However, non-linearity in the 'fluid flow' of the universe is so slight within currently measurable distance accuracies it is questionable if the difference can be detected. The potato analogy is of course only a one dimension flow explanation, but since galactic decelerations are in three dimensional space, directional sensitivity would apply in all three dimensions (six directions). In directions perpendicular to the main flow there would be growing separation, but not as significant as in the flow line direction. Other common analogies (in reference) relate to two and three dimensional processes which further demonstrate the plausibility of this concept for origin of the universe.

<u>Two reasons why accurate measurements of distance and red-shift observations of some galaxy cluster's recession velocities are irregular:</u>

1. Almost all galaxies are members of groups and clusters. Galaxies are circulating (revolving) about other galaxies as well as clusters revolving about other clusters. Galaxy groups are receding, but individual galaxies' circulation in various directions adds or subtracts an intermittent increment of red-shift for the measured 'cluster' recession rates.

2. There is a measurable difference in red-shift (rate of separation and deceleration) due to our off-set location from the site of origin of the Deflagration wave. Of course this assumes we are adequately separated from the site of origin that non-linear deceleration will reveal the direction to the center. If we can make red-shift and distance observations of galaxies on the other side of the origin, the asymmetry might be revealed. (See comments on voids). This will eventually be measured and could identify the asymmetry we want to detect and eventually measure.

Laws of Physics requires that momentum must be conserved; deceleration is caused by linear momentum being surrendered / transferred to angular momentum through the processes of mixing and vortexing. The mix rate is expected to be non-linear, like the pressure drop in all fluid flows is a second order 'square' function with distance. ($\Delta P = k*(\rho\ v^2)/2$), [12, 13]. Thus, the linear velocity will decrease with distance traveled as the mixing and angular momentums increase. The curvature of asymmetry in the universe is expected to be nearly flat because of the tremendous distances involved; therefore it may not be directly measurable with current technology. Linear decelerations of groups and clusters are caused by interaction between, within, and among galactic clusters.

When specific galaxies are observed repeatedly over a sufficient number of decades, observations will prove asymmetry; and it will then be undeniable that galactic clusters are in fact decelerating apart. There continues to be decelerating growth occurring on our side of the outward advancing deflagration front. Outward linear velocities will continue to be consumed behind the deflagration front via galactic clustering and vortexing. Deceleration stream lines are illustrated and labeled in Figure 10.1 of Reference [1], and also in the following figure 4. Two Hubble lines are superimposed over deceleration lines and reveal H# lines to be gradients across the lines of deceleration. Continuing growth of the universe means farther separation of galaxies with time; growth will continue 'ad infinitum', even if the primordial matter becomes depleted.

Universe Growth

Close behind the D wave there are tremendous amounts of new orbiting, coalescence, and other transfers from linear momentum. More red-shift vs distance indicates decelerations are slowing.

Figure #4

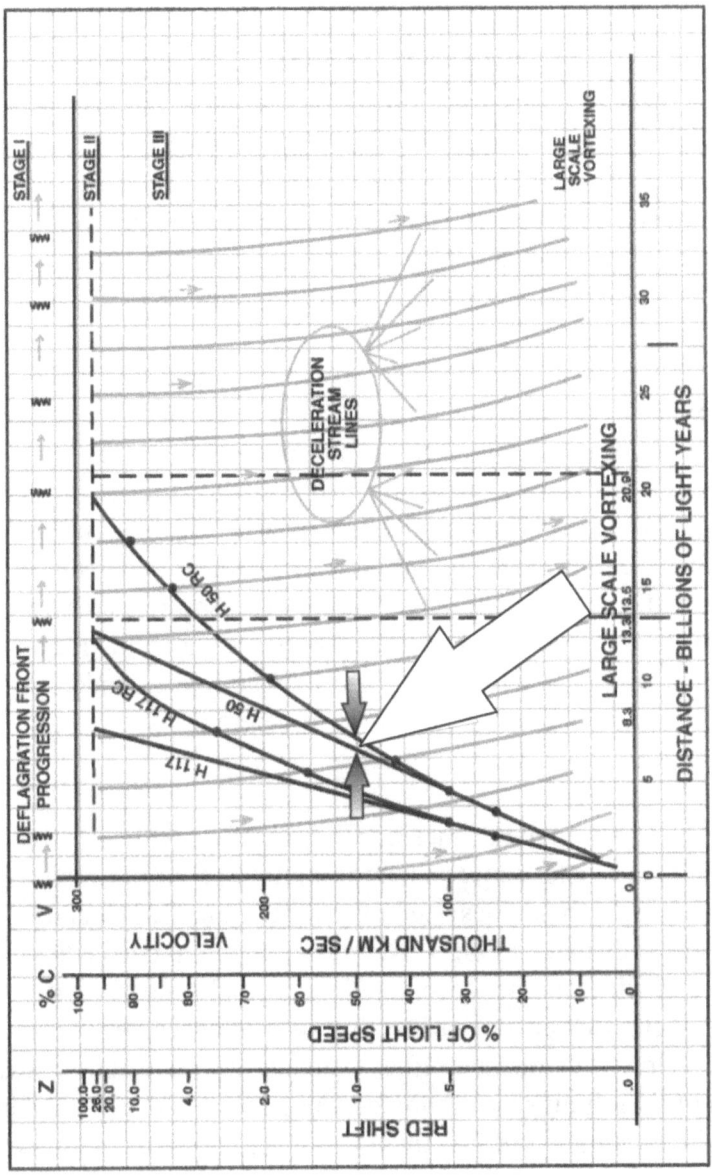

Observable Universe diameter & Our distance to the center

It is assumed the universe is spherically symmetrical about the origin (location of the first positronium annihilation that started the chain reaction) of the deflagration wave that produces all of the corporeal matter in the universe. Evaluating the asymmetry from our location is expected to define how far we are from the universe's origin. Estimating our location in the universe was first attempted [1] by comparison of modern times Hubble number measurements (Between years 1980 and 2000) by knowledgeable, reputable astronomers. The largest H# (117) represents higher velocities vs distances at a given direction and the smallest H# (50), represents the lower velocities vs distances in its direction.

Since the highest and lowest H#s were measured in opposite (approximately) directions, they were expected to reveal the asymmetry of the universe from our vantage point. The largest H# is towards the closest proximity of the deflagration wave and the smaller H# is towards the farthest distance to the deflagration wave; therefore the difference should indicate different deceleration rates. Graphing these Hubble lines ([1] Figure 10.3) shows us where we are relative to the center of the universe. Since the deflagration wave travels at the speed of light (c), the Hubble line extended to 'c' on the graph should show the distance to where the deflagration wave is, or appears to be, today.

That was a first estimate, but since the age of white dwarf stars in the M4 globular cluster show us the minimum age for our region of the universe, the Hubble lines need to now be reconciled (first iteration). The closest distance to the wave was assumed to correlate with the steepest H# line, and was reconciled accordingly; the farthest distance to the wave (lower slope) H# line was adjusted proportionately. These velocity vs distance lines extended to intersection with the speed of light limit show us the distance to the deflagration wave in opposite directions. Adding these distances shows the diameter, and the difference shows our distance to the universe's center ([1]; Figure 10.1a, Figure 10.3).

Since Hubble lines are gradient lines, they should approach the deflagration wave almost asymptotically at the speed of ~97 % of c, but become tangent at the point where the universe's age corresponds to ~30 billion light years distance from the origin. Direct measurements of divergence from straight line projection of Hubble #s, i.e. distance vs red-shift, are being studied.

Supernovae (Sn1a) provide distance markers at red shifts to, and above, Z = 1: (Reference [26]). (Z = 1.0 corresponds to recessional velocity at 50% of light speed) Calibrating the Hubble line curvature is now being achieved with Sn1a observations.

The intrinsic brightness of all Type Sn1a supernovae, are reasonably consistent. They are so bright they outshine their entire host galaxy for a few weeks. By observing their apparent maximum brightness their distance can be calculated and the recessional velocity can be accurately measured by their red-shift. Recent studies indicate brightness calculated distances can be further refined by measurement of nickel content of the novae emissions, allowing more accurate distance calculations.

Recent (2006, 2007, 2008) Type Sn1a supernovae red-shift and brightness (distance) data confirm the NUT predictions are valid for gradient (H#) lines for red-shifts to about Z = 1 (50% of light speed). Albeit astronomers' motivation for obtaining supernovae data is for another purpose; BB theory dark energy researchers hope to discover something that might reveal a mystery and support 'hypothetical and mythical dark energy'. Contrary to the BB, but complementary to this 'New' universe model, non-linearity along mono-directional measurements substantiates deceleration flow characteristics.

Valid (off axis from origin) asymmetry can be quantified only when Sn1a red-shift and apparent brightness data are correlated with direction. The origin of the universe (NUT) concept presumes it is highly likely that we are not at the center of the universe; otherwise we would be in or adjacent to a large void. The universe is growing, but not expanding (there is a major difference). Asymmetry (different H#s in different directions) as well as non-linearity of Hubble lines (Indicated by opposing arrows on Figure 10.1a) are expected to be more obvious at greater distances, where red-shifts are significantly above $Z = 1.0$.

Pre-1990 H#'s value variations by astronomers (although probably measured accurately) are due not only because of different directions but also due to the vorticity movements of galaxies in and among the nearby galactic groups and super clusters. On the other hand, the deviation of the line through recent H#s at $Z = \sim 1.0$ from the H# straight line is strong evidence that the curved H# line is a deceleration gradient line.

Measured divergence of Hubble number data from the original Hubble assumed expansion straight lines should be recognized to not be the result of present or past accelerations. However, continued use of Hubble numbers for distance estimates from red-shift measurements between 0.1 and $\sim < 1.0$ provides a convenient yardstick, but distances so calculated must be recognized as approximate. When adequate data has been obtained of Sn 1a red-shift and distance by specific direction, and in all directions with sufficient detail to define the equations for each directional Hubble number line, red-shift and direction with the yet to be empirically developed applicable equations will provide us with an accurate yardsticks for each direction.

The following graph is a reverse / mirror image of the Sn1a data plot from a 'dark energy research team' of astronomers, as reported in 2007. Their observation results precisely reveal predictions by this "Model of the Universe with Laws of Physics". 'Dark energy' is discarded in favor of 'deceleration' interpretation.

Figure #5

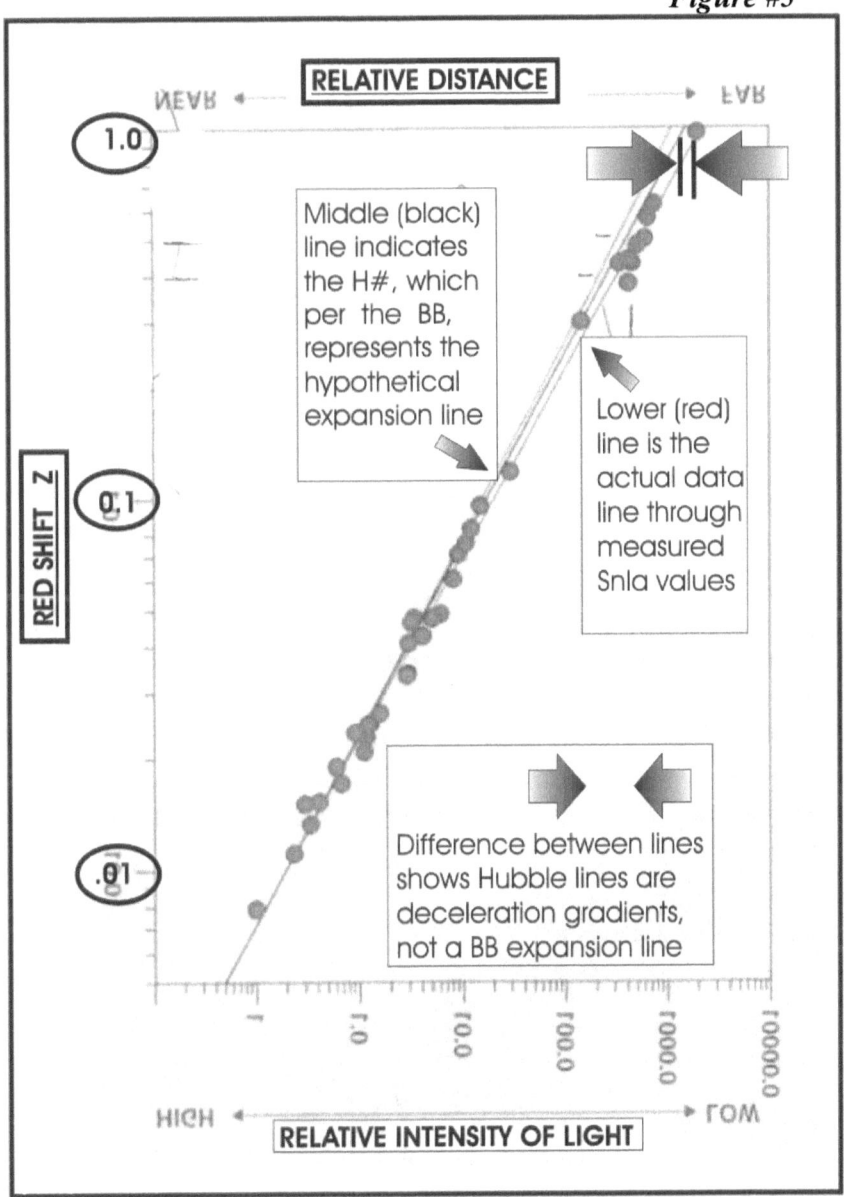

Following are the 'dark energy' study team's published data
from which the mirror image graph was illustrated in Figure 4
and Figure 5.

Figure #6

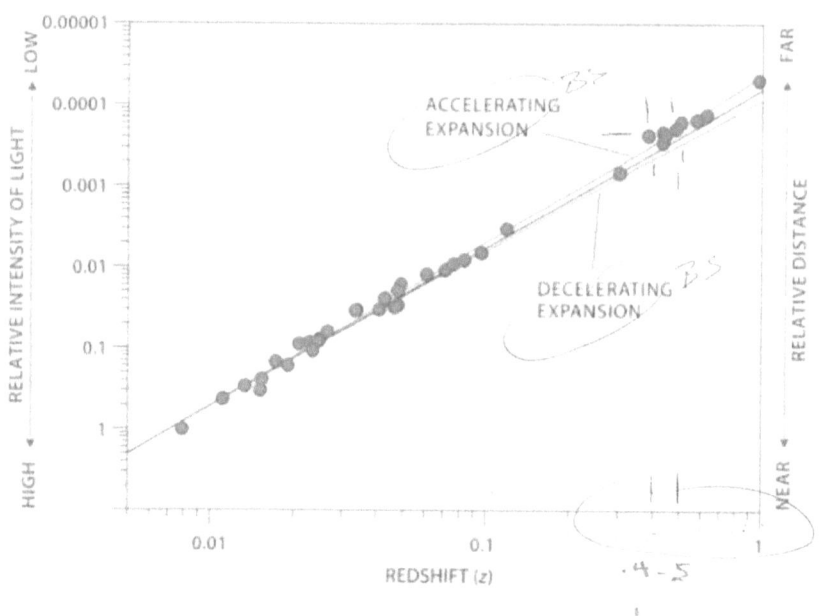

*The NUT Model of the Universe predicted Hubble line nonlinearity
is emphasized by the opposing arrows on Figure 4. The predicted
deviation from a straight line is now confirmed by the opposing
arrows on the data on Figure 5, as acquired and presented by the
'Supernovae Type Sn1a' 'Dark energy' research group, Figure 6..*

Further substantiation: ... Data as plotted by the Adam G. Riess 'Dark energy study team' substantiates the NUT origin of the Universe Model concept

(Arrows added to emphasize Red-shift Z > 1.0 to Z = 1.755)

Figure #7

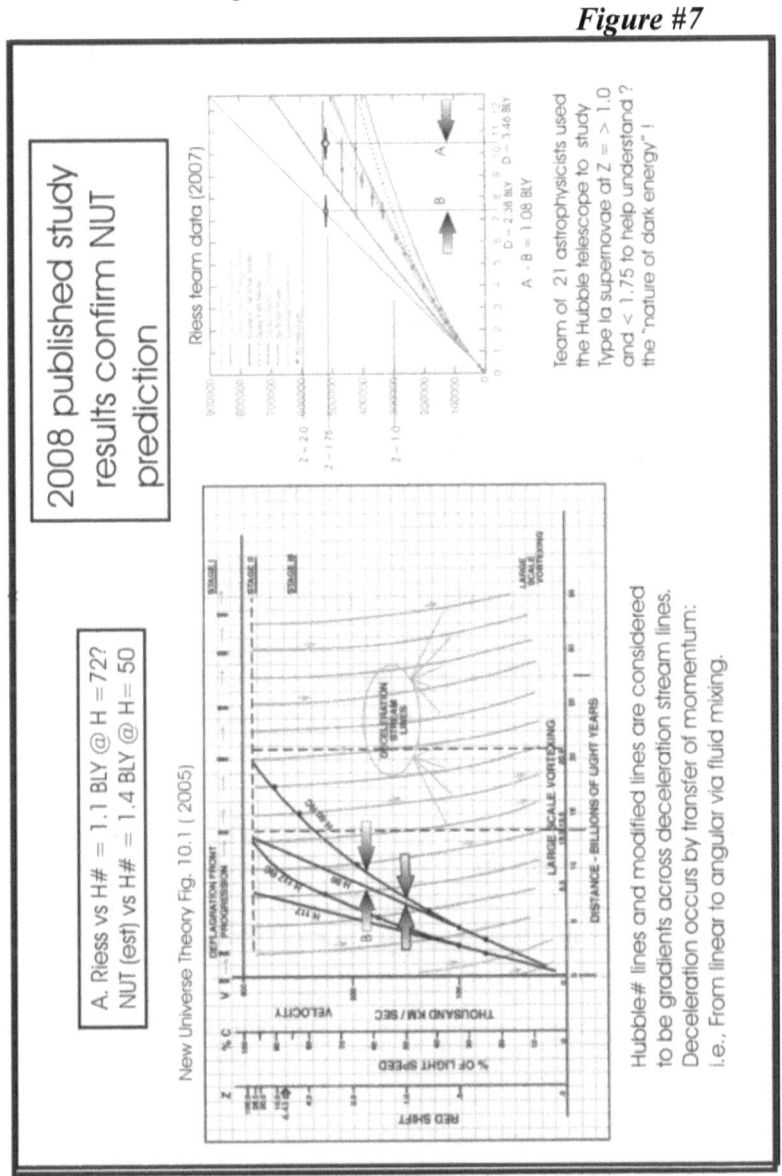

Breaking News! **Study results including Sn1a at Z = 1.755, becomes aware to me in March 2008, 1st published in 2007.** **Supernovae Type 1a red-shift data; Z > 1.0 to Z = 1.755 [7].**

Recently obtained data demonstrating deviation of distance vs red-shift (from straight line) has been reported in late 2007 and published in early 2008 in the Astrophysical Journal. The data is available through 'arXiv.org,' a web site operated by Cornel University. These data further substantiate this Model of the Universe, and the NUT and the following two graphs show the data impact on 'straight' H# lines. The data is from the observations by an elite/special team of astrophysicists researching the nature of 'dark energy?'!

The team includes 21(?) astrophysicists, and is headed by Adam G. Reiss (JHU, STScI). The team is working under the handicap/ constraints of the presumption of the BB and an expanding universe. Their data graph follows and their data has been included on the following copy of Figure 10.1 from Reference [1], with an additional pair of opposing arrows.

The curved Hubble lines were estimated in 2003 when the NUT was originally being developed and written. Hubble lines were recognized as deceleration gradients. Actual / measured distance data at high red-shifts (~ 1 & > 1) were not available at that time.

The NUT predictions are confirmed, and the evidence becomes overwhelming and compelling that this plausible 'Origin of the Universe' concept is valid. It is also compliant with all Laws of Physics and is now referred to as the New Universe Model.

Voids

Significance and origin: Voids strongly support the NUT 'Origin of the Universe' concept which provides a likely explanation for their existence. However, the known void locations are puzzling. They are believed to be fragments of the original void but some are in directions which are contrary to the asymmetry demonstrated by Abell (ACO) cluster study. (See 3. Angular Momentum, page 32)

The location of the start site of the deflagration front from the initial positronium annihilation is identified on Figure 10.3 [1]. A large void is expected at the center of the Universe; labeled 'Origin X Void?' A large spherical void, with about 4 to 8 billion light years original diameter, was expected at the origin site because the front would require one to two billion years for the deflagration front to transcend an adequate number of positroniums, precipitate outward flowing mass particles and objects to accrete into stars, and then continue their outward decelerations by turbulent mixing of galactic clusters.

The question mark on the figure was to emphasize it was not known whether the original void would still exist or have been eroded into fragments after the past 25 to 30+ billion years. It was / is not known how long will be required for turbulence and vortex mixing to fragment, erode and 'back fill' that void. Void fragments are expected in accordance with the 2nd Law of Thermodynamics; and as stated by Alan Shapiro*: "Old vortices never die, they just 'get larger' and fade away." References [8,9] (* often referred to by many as the father of modern thermodynamics)

A huge void, almost half a billion light years in diameter and about a billion light years away, was discovered in Bootes in 1981 by astronomers Robert Kirshner, Augustus Oemier Jr., Paul Schechter and Stephen Shectman. They reconfirmed the void finding, after further surveying the area in 1983. By 1997, 60 galaxies had been found in the Bootes Void; they are mostly located near the void edges as would be expected by erosion.

The 'Bootes Void' is the largest of several voids known prior to the turn of the millennium. It is 'only' about .300 billion light years in diameter, and therefore it is thought to probably be only one of several remnant fragments of the original central Void. The Bootes Super Clusters are combinations of large galactic clusters that lie beyond the Bootes Void. Finding a Sn 1a in a few, or even one, of the Super Clusters beyond the Bootes Void might reveal significant or even conclusive asymmetrical evidence. (e.g., in Abell's #'s A1861 and A1795)

Discovery of the largest void found to date, (published in August 2007 by Lawrence Rudnick and colleagues of University of Minnesota, Reference [12]), is possibly the largest still in existence. It is over 1+ billion light years in diameter, and is 6 to 10 billion light years away. (At a red-shift of about 1.0) This void is located in the southern hemisphere in the direction of constellation Eridananus, (R.A. ~3hr 30 min, ~Dec -50 degrees). This void is named as a cold spot, since there is very little energy in that region. Such a radiation free volume is also expected for the original annihilation site in primordial matter, as all radiation and the chain reaction deflagration wave progressed and continues in all directions away from that start location. This region is referred to in Reference [1] Figure 7.7 as CENTER OF UNIVERSE, and in Figure 10.3 as ORIGIN X VOID. The distance from us is about twice the estimate shown in the NUT book, but the concept is validated.

The direction of the Bootes Void is in the northern hemisphere, (R.A 14 hr 30 min, Dec ~ + 45). The directions to these two separate voids are in approximately opposite directions from us and are separated by billions of light years in distance. If these and other voids were all once part of the same original central void it is indeed hard to visualize how even the universe's large scale vortexing could result in such wide spread void fragmentation and separation. But it had a long time to occur, and apparently did happen in the ensuing ~ 27 billion years following the initial positronium annihilation.

Decelerations of other galaxies and smaller voids on our side of the largest known to date (original / central?) void are logically expected to be traveling in the same direction that we are decelerating. Decelerations of galaxies and galactic clusters on the other/far side of the original central void will be receding in the opposite direction. Therefore, we might expect to find a measurable difference in Hubble number values on the distant, opposite, side of the central void, if we knew where that was. This is definitely an area of study that deserves some significant research efforts.

Erosion and fragmenting of the original void could have produced confusing H# results, in the vicinity of the void fragments. Obtaining red-shift and apparent brightness of Sn1a supernovae at these distances in the proximity of the voids will possibly give us some clues. Void study may not receive high enough priority for some time in the future. It could help to get the Astronomy community and especially to get education of future astronomers on a foundation of scientific thinking. In other words, let's form opinions with proven facts and using scientific thinking. Of course speculation is needed to define possible optional answers to be investigated.

Many who review this Model of the Universe have speculated about the origin of multiple voids? The speculative answers are:
1. Erosion and fragmentation of the central void at the initial annihilation site;
2. Multiple early annihilations with each having their own void;
3. A statistical anomaly or chance alignment of galactic clusters.

There are reasonable arguments for each explanation, but the question will eventually be answered by analyses accomplished by astronomers of the future who will be equipped with more observations. Mythical explanations cannot be proved valid or invalid.

Section view of Universe: Note central VOID prediction (Reference [1] Figure 10.3)

Figure #8

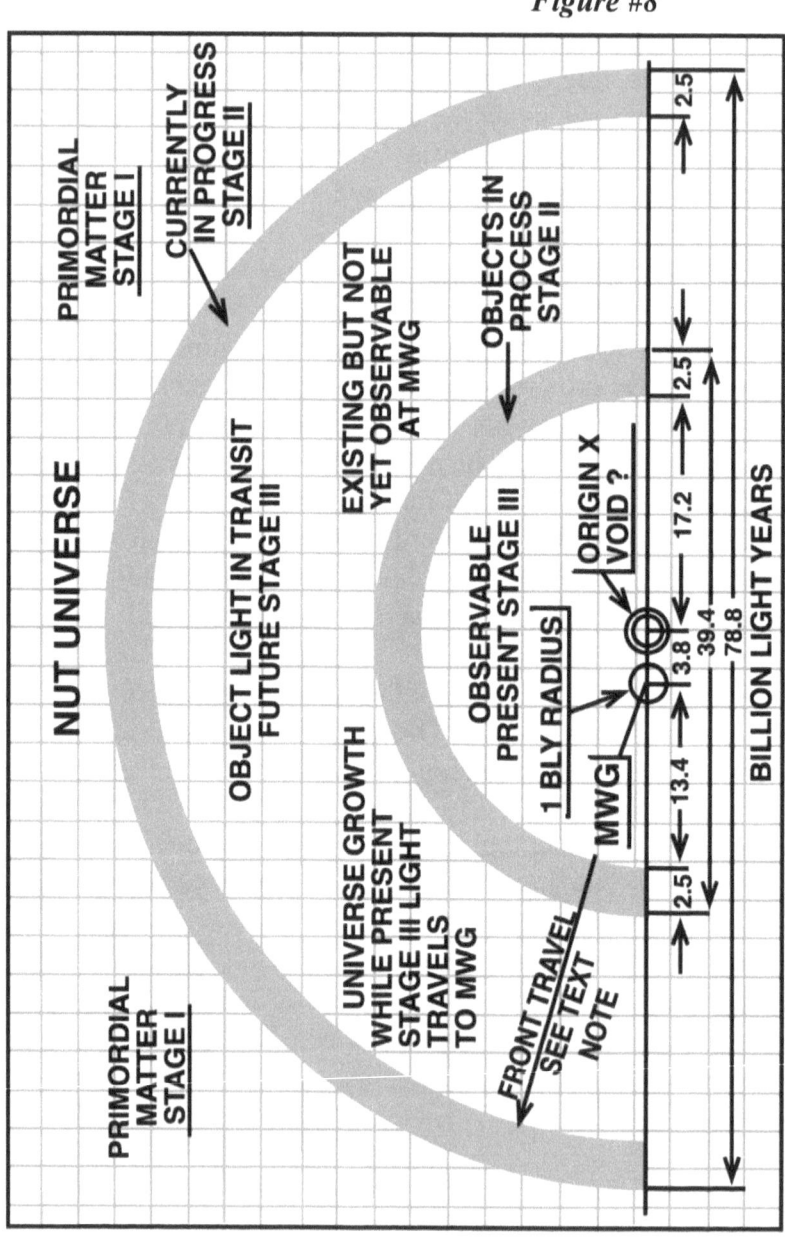

Quasar observations:

Accurate distance measurements are desperately needed for some of these high Z (red-shift) quasars so that the gradient line of figure 20, (page 62, fat arrow #2) can be calibrated at Zs of ~ 6.0+0.

Dr Jack Mitton, RAS [10] reported in April 2003, a UK team of astronomers (including Dr Kate Isaak of Clavendish laboratory at U of Cambridge, Dr Robert Priddey of Imperial College in London, and Dr Richard McMahon U of Cambridge) using the James Clerk Maxwell telescope in Hawaii, measured the red shift of 6.43 for the quasar SDSS J1148 + 5251 in the direction of Ursa Major.

In 2006, the UK team measured the highest ever measured red shift for their most distant quasar that corresponds to the recessional velocity of approximately 87% of light speed I. It is theorized in the new universe model, red shifts up to $z = 27$ (97% of c) will be the highest red shift observable. In accordance with the NUT definition of the mass outflow from the deflagration wave (behind the outward progressing wave), ~97% of c is the velocity at which the first generation stars ignite their thermonuclear furnaces to start radiating light. Examination and comparisons of Reference [1] Figures 10.1 and 10.3 illustrate how the UK team red shift information agrees with the location of the deflagration wave at the time when light started from that distance.

The deflagration wave is now at least twice that distance away, and light from the wave's current distant position won't reach our region of the universe for another dozen or two billion years in the future.

Calibration of red-shift vs distance at such large space separation is a problem yet to be resolved. Who will be the first astrophysicist to discover a technique to measure such distances; Of course it will not be using H#s.

3. Angular Momentum; Vortexing and Galactic Clusters

Large Scale Vortexing and Momentum Conservation

What is the source of the tremendous amount of energy in the angular momentum in the universe?

Thousands of billions of galaxies in the universe are all postulated to have originated from a point source according to proponents for the BB theory. All mass objects are particle combinations that had to form from divergent traveling particles while they continued to radiate outward in all directions. Divergent (BB) linear momentums would inhibit, if not prohibit, particle fusions, star and galaxy formation.

On the other hand, as in the deflagration wave model concept, (NUT), particles precipitate into a radiating circumferential fog at near the speed of light. The 'fog bank' of particles have extremely high enhanced gravities. With velocity enhanced gravitational coalescence, the fog fragments into clouds; particles fuse and further coalesce and clump within fragmenting clouds. Since <u>none</u> of the particles coalesce and come together perfectly 'center-of-gravity' to 'center-of-gravity', the resultant particle combinations each acquire varying amounts of angular momentum. With each encounter linear motion energy is extracted from their outward momentums. Deceleration occurs incrementally and continuously as momentum is conserved.

Recognizing this process is repeated trillions of trillions of times within each micro-steradian, an understandable and plausible explanation for the energy source for all of the angular momentum in the universe is the deflagration wave. This also provides a plausible explanation for increased red shift with distance while the objects are decelerating outward from and behind the outward progressing deflagration wave.

Eureka! ...4,076 ACO Trombones

At the 2007 annual joint meeting of the AAS and the AAPT in Seattle, I explained this new concept to a bright – curious young Professional Astronomer. Matthew Shetrone is from the University of Texas' MacDonald Observatory. He was receptive to the NUT idea. However, Matt recommended that to provide a convincing argument, a way needs to be found to demonstrate asymmetry, which would maybe verify the new concept and nullify the BB. He suggested use of Abell clusters.

Part of the basic argument for the NUT model is that the increasing red-shift verses distance 'expansion' is a series of measurements of <u>different</u> individual galaxy velocities, and as such different object measurements do <u>not define</u> acceleration. To confirm acceleration or deceleration, <u>there must be at least two time-separated red-shift measurements of the same galaxy over a significant time period</u>. 'Consistent with Laws of Physics NUT' recognizes that galaxies can be decelerating by conversion of linear momentum to angular momentum (increasing vorticity via clustering). Acceleration would require a hypothetical force for each particle. On the other hand, humongous quantities of angular momentum exists in the form of rotating multiple star systems, in galaxies, and in galaxy clusters which continue to increase through compounding growth of galactic clusters and super-clusters.

George Abell (1927-1958) of Cal Tech cataloged 2,712 clusters each of which include 30 or more galaxies from the PASS photographic plates. Harold G Corwin of Cal Tech, formerly University of Edinburg, and Ronald Olowin of University of Oklahoma, completed the ACO project to total 4,076 clusters which includes the southern hemisphere. (PASS was the Palomar All Sky Survey.) They extracted and documented data for galaxy-clusters with counts of over 30 galaxies in each. This cluster catalog is now known as the **ACO** catalog.

The ACO / PASS sky survey data has maximum red-shift (Z) values up to .318, which corresponds to distances of about 3.5 BLY from the Milky Way Galaxy (MWG). Since the ACO data was compiled, larger super-clusters (Groups of mutually orbiting galaxy-clusters) have been observed at greater distances with counts in the thousands.

Analyses of the ACO data reveals exactly what is expected from the universe if it formed from primordial matter as described by the New Universe Model. Older regions of the universe will contain larger galaxy count clusters (i.e., larger vortices, just as occurs in all other typical fluid flows). ACO galaxy-cluster data sorting has been done to identify the largest 15 and to identify their direction, since the larger ones direct us towards the origin (oldest region) of the Universe. Smaller clusters are in almost all directions from the origin and from us (the MWG), since we are also in a small galaxy cluster of less than 30 count. If we had a catalog of all lower count galaxy clusters, to maybe galaxy counts as low as 10 or 15, we might observe quantity directional asymmetry for smaller clusters.

The largest galaxy counts in galaxy-clusters are where the galaxies and galaxy clusters have been in existence, accreting, vortexing, and decelerating for the longest time. Therefore their location must be in the direction towards the older, central universe region.

Allen Shapiro's quote is worth restating; his quote: "Old vortices never die; they just get larger and fade away". The same principle applies here as linear momentum of galaxies in the extra-galactic medium has no way to dissipate except by more compounding with other angular momentums, ad infinitum.

The direction towards the deflagration wave is towards the larger population of smaller count galaxy clusters, which have not yet had the time to accrete into larger count clusters. Of course, this applies everywhere in the universe, not just on our side of the universe's center/origin. Single galaxies and small count galaxy clusters are closer to the front, in the younger regions, which are also still traveling faster with their higher linear velocities.

The average direction was calculated for the 30 lowest count ACO clusters. Results were non-conclusive as expected; our Local Group is in the midst of 30 count clusters. Also, the 30 count clusters are predominately in low red-shift of .02 to .08 while the 15 highest 30 count Z's are mostly .1 to maximum of .29. (Red-Shift is an accurate measure of <u>only</u> the radial component of velocity.)

The highest 15 galaxy count clusters out of the 4,076 ACO Galaxy Clusters, are graphed to reveal the direction towards the center and origin of the universe. Our motion and deceleration is in the opposite and outward direction.

(Highest 15 count ACO CDS data are included in appendix)
(*ACO Celestial coordinate version 1950.0*)

After graphing the 15 highest count galactic clusters, the J2000.0 Sky Atlas [34] was consulted; the highest galaxy population direction is revealed and, as illustrated, is within about one hour RA of the ACO origin direction, and within about five degrees Dec.

Red Shift and spectrograph instrument resolution for the sky atlas data is elusive and unknown to me.

The 15 highest count ACO galactic clusters

Large arrow points towards the largest, oldest, galactic cluster region.

Figure #9

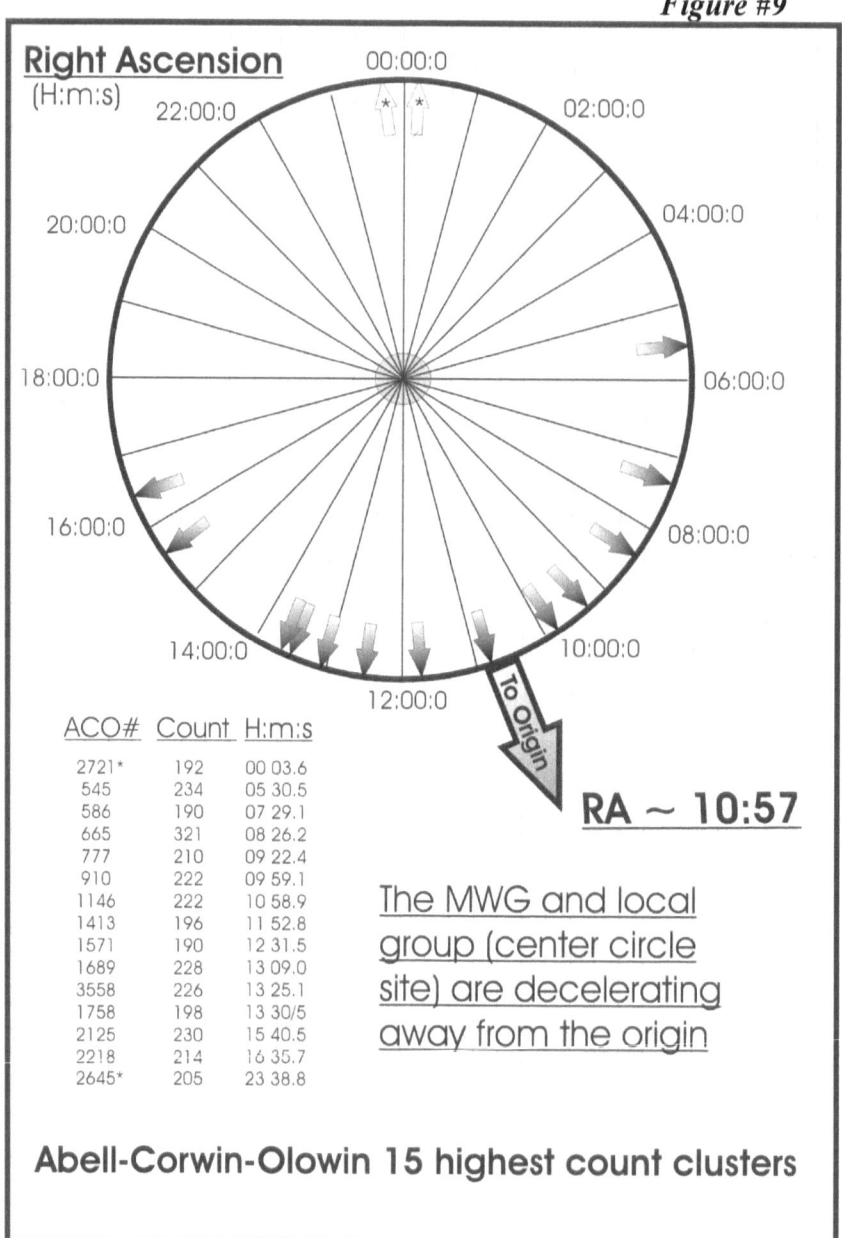

Right Ascension (H:m:s)

ACO#	Count	H:m:s
2721*	192	00 03.6
545	234	05 30.5
586	190	07 29.1
665	321	08 26.2
777	210	09 22.4
910	222	09 59.1
1146	222	10 58.9
1413	196	11 52.8
1571	190	12 31.5
1689	228	13 09.0
3558	226	13 25.1
1758	198	13 30/5
2125	230	15 40.5
2218	214	16 35.7
2645*	205	23 38.8

RA ~ 10:57

The MWG and local group (center circle site) are decelerating away from the origin

Abell-Corwin-Olowin 15 highest count clusters

28

Large arrow points towards the largest oldest galactic cluster region.

Figure #10

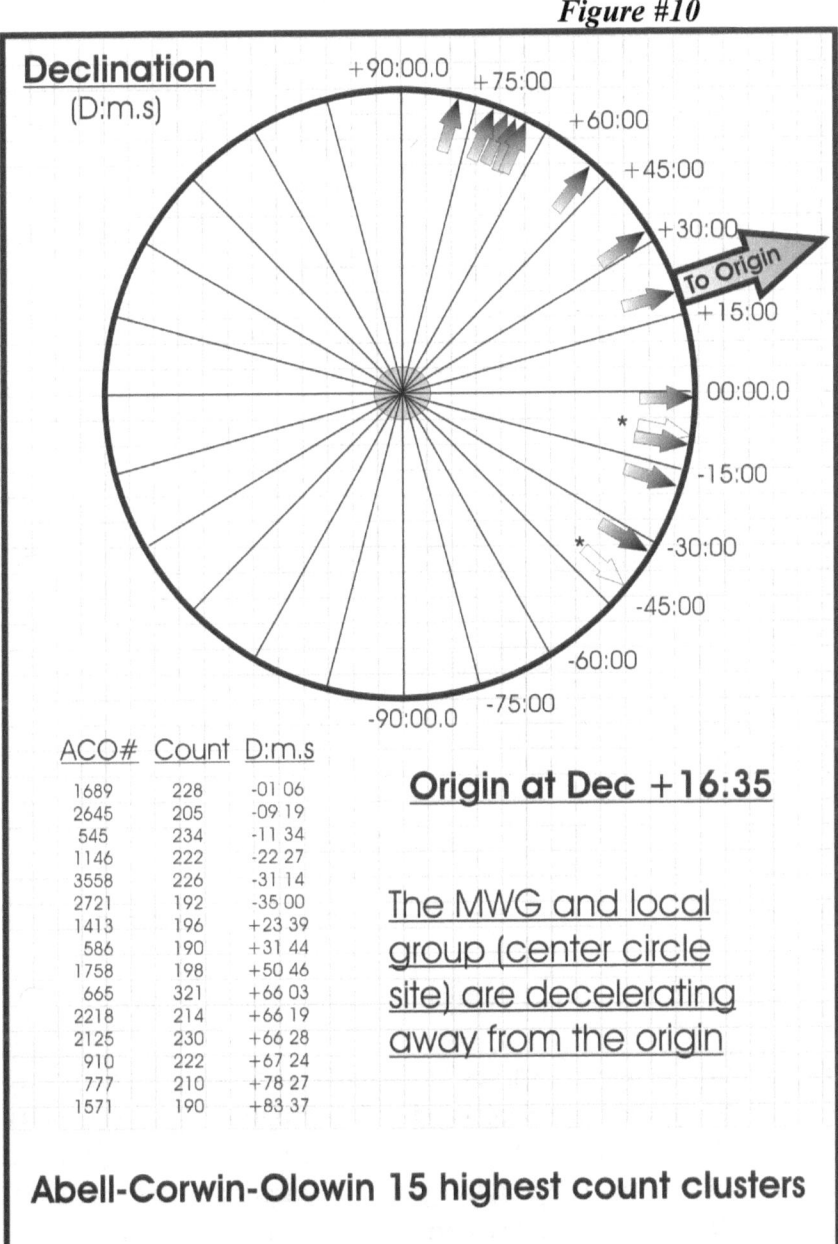

Declination
(D:m.s)

ACO# Count D:m.s

ACO#	Count	D:m.s
1689	228	-01 06
2645	205	-09 19
545	234	-11 34
1146	222	-22 27
3558	226	-31 14
2721	192	-35 00
1413	196	+23 39
586	190	+31 44
1758	198	+50 46
665	321	+66 03
2218	214	+66 19
2125	230	+66 28
910	222	+67 24
777	210	+78 27
1571	190	+83 37

Origin at Dec +16:35

The MWG and local group (center circle site) are decelerating away from the origin

Abell-Corwin-Olowin 15 highest count clusters

Asymmetry, real and beyond doubt !

The data compilation by ACO team have only a few peculiar anomalies, and I did not want to edit (alter or taint) the original listings by ACO contributors, so all ACO data was used as cataloged. Note; Milky Way Galaxy and Local Group cluster (MWG & LG) are shown at approximate location.

Figure #11

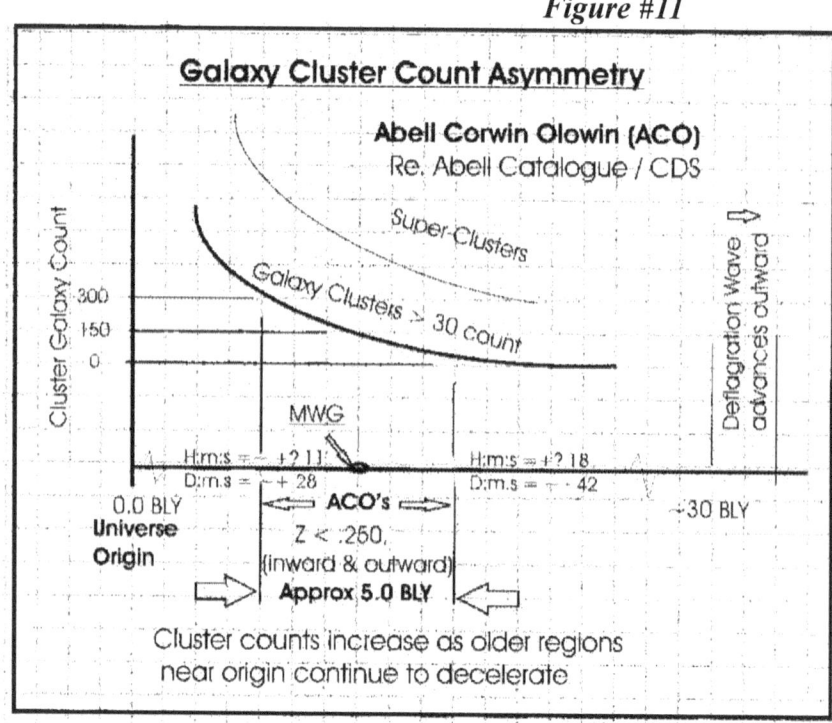

(Also, see ACO Population Distribution Graph)

The WMAP asymmetry survey of the universe illustrates asymmetry for radiation, whether background or foreground. ACO asymmetry data has been available for decades. Thanks Abell, Corwin, and Olowin. In this ACO population graph, the larger clusters (vortexes) are to the left which indicates the direction towards the older universe. The universe is therefore growing to the right in this view.

In 1957 George Abell (1927 – 1958), used photographic plates (from POSS, a Palomar Observatory Sky Study) and compiled a list of 2,712 galaxy clusters with red shifts up to Z = ~ 0.24, each containing at least 30 galaxies, including some with over 300. In 1989 the list was extended to 4,076 by Harold G Corwin from the University of Edinburgh (now at Cal Tech), and Ronald P Olowin of Oklahoma University. This completed the ACO catalog which includes the Southern hemisphere. (as possible from Mt Palomar)

Figure #12

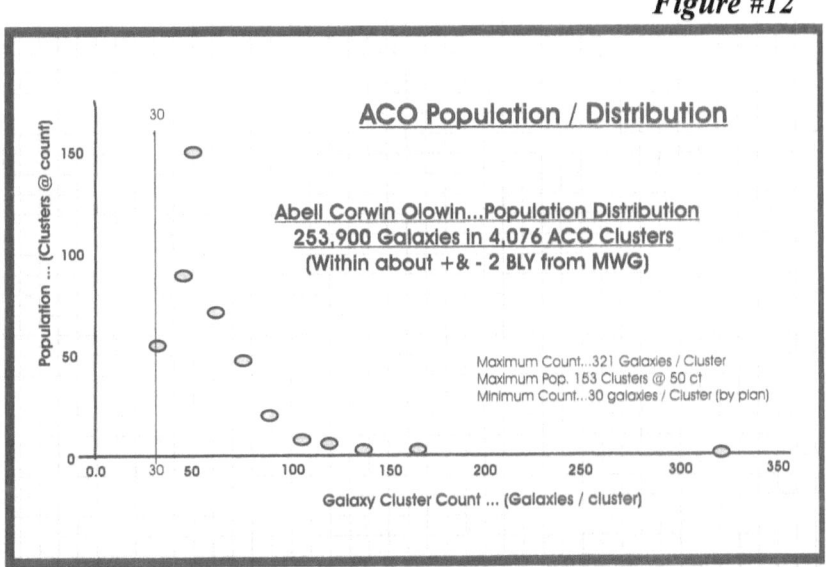

Largest and therefore the older clusters are to the right in this graph. This asymmetry proves the universe is growing in the opposite direction (to the left); towards, and following the deflagration front that is precipitating decelerating objects. Why the total ACO Population / Cluster Count peaks at 153 clusters at 50 Count is not yet understood. Our 'Local Group' is to the left of the 30 Galaxy Cluster Count Line. *(The distance span for all ACO data is less than about 4.5 BLY).*

Follow-on studies suggested for undeniable proof the universe is growing, but not accelerating or 'expanding':

Measure the red shift of specific ACO galaxies over time lapse intervals, of say several decades or more years.

This will demonstrate if 'original' ACO cluster catalog red shifts are constant, not increasing. That would further show no higher velocity now than when the previous velocity (red-shift) was acquired, even though the galaxy brightness will prove the same galaxies are now farther away. After more decades ACO 'Z' (red shifts), will decrease with time and it will become apparent and proof the deflagration wave products (galaxies) are decelerating. This will disprove the presumption that all matter originated from a single point.

Red-shifts of some of the ACO clusters were obtained by measuring the red shift of an alpha galaxy (brightest in the cluster). Clusters that did not have a readily identifiable alpha galaxy were catalogued from an average of two or more galaxy red-shifts. The single alpha galaxy red shift catalogued clusters should be selected for time lapse study to minimize uncertainties. There will be some alpha galaxy red-shifts increasing and some decreasing due to circulation of the galaxies within the clusters as the alpha galaxy is not likely to all be located precisely at the center of gravity of the total cluster. Also, the cluster center of gravity will move about as galaxies circulate within their cluster systems.

I mentioned this suggested study to Dr Harold G Corwin Jr., (of the three original ACO researchers and cataloguers); He informs me that a 1999 Database [38], by Mitchell F Struble and Herbert J Rood (S&R) ApJS 125, 35 contains recent and ACO red-shift data. Corwin suggested the low red shift data available may not be adequate to answer acceleration + or -. He informed me how to access the data through NED [42].

All 583 S&R cluster records cataloged in 1999 that pertain to same clusters as some of the 4076 original ACO clusters which have red shift data have been reviewed. The 4076 ACO clusters were catalogued in at least two groups: The 1st group of 4076 by Abell in 1957, and in 1989 the 2nd group of 1364 clusters by Corwin & Olowin.

The significance of the 583 S&R clusters is they have two sets of time lapse red shift data acquired decades apart! 228 of these had zero red shift change (Listed in figures 23 & 24), some for four decades and all for at least one decade before 1999 … indicating traveling at constant speed with no acceleration or deceleration.

Conclusion: No expansion?!

But to answer the question; "Are a few decades of time long enough to prove no acceleration?" Analysis of these data indicates the answer is YES! Based on the Hubble 'equation' the red shifts would have increased by a significant amount. Calculations strongly indicate the universe is not expanding.

In keeping with my practice of not mixing equations with text; proof the universe is not expanding with acceleration is included in the appendix. Analysis of the highest red shift in the 288 is presented in Appendix 2, page 74. Pertinent equations and data are listed in Appendix 6, page 87. **

Some of the S&R's 355 ACO cluster's alpha galaxies were at higher red shifts, and some had lower red shifts. Obviously, these non-zero change red shift alpha galaxies were not at the gravitational center of their host cluster. All of the ACO cluster's galaxies meander throughout and within their host cluster. They are dynamic, confined and guided by their momentums as well as their interacting gravitational fields.

** If you try some of the Appendix 2 calculations (and I hope you do), it is essential your calculator can handle at least 21 significant figures. Calculators TI-30Xa and a HP equivalent lost digits in calculations. Consistent results were obtained from a TI-84-Plus-Silver.

Angular Momentum Analyses

I occasionally ponder the task of estimating the total angular momentum in the universe in an attempt to correlate and estimate the mean deceleration rates for clusters of galaxies in the growing universe. The task would be monumental. However, as reported in July 2007 Science, internet help has been solicited which will provide some basic data.

Astronomers at Portsmouth and Oxford in UK and at John Hopkins in the US are requesting support for their study of galaxy rotational statistics. They are using photographic data from Sloan Digital Sky Survey (SDSS) obtained from the telescope at Apache Point Observatory in Sunspot, NM; (Oxford astrophysicists Kevin Schawinski at Oxford and Bob Nichol at Portsmouth). They have named the effort Galaxy Zoo, (www.galaxyzoo.org). Reference [27]

They have solicited help from any and all internet computer users; after recruits 'sign up' they are trained to recognize various galaxy types and rotations. After a simple test, the volunteers are given a group of catalogued galaxy pictures to analyze. Results from the volunteers are collected in a data base which will be processed by project astronomers.

As reported in the web site for Galaxy Zoo | The Science | Project |: "Professor Michael Longo from the University of Michigan has suggested in his recent astro-physics preprint, that there may be a preferred 'handedness' (rotation direction) of galaxies in the local universe. He has stated "This is a revolutionary claim that could force us to rethink our understanding about the underlying nature of space and employ a much more complicated background model for the universe." Their initial claim is based on a sample of just 1660 galaxies from SDSS survey, but a much larger sample is required to assess the significance of the effect, which is where 'Galaxy Zoo' comes in." An additional study 'Galaxy Zoo 2' was started (in 2009).

Consistent with and supporting this New Universe Model; there are no known sources for angular momentum other than by transfer from and reduction of linear momentum. Momentum transfers occur as orbiting, a consequence of mutual attraction forces between nearby objects resulting in linear deceleration of the combination.

Estimating the <u>total</u> angular momentum in the universe will require a complex systems analysis, and a lot of sophisticated educated estimates. To be considered are sub-atomic rotations as well as the multiple star systems in mutual orbits (binary, tertiary, quaternary, etc), galaxy rotations, galactic black hole core consumptions, galactic cluster mutual circulatory rotations, galaxy mergers and continuing mergers that occur between peripheral galaxies in adjacent revolving galactic super clusters. These are all various forms of angular momentum increases which will vary with time and distance as accretions and accumulations traverse space. In any total momentum analyses, to be considered are the many mergings and collisions of various sizes of conglomerates which result in some of the momentums being converted into heat and entropy.

Total momentum analysis of the universe will be a humongous and monumental task but maybe not an impossible one, and in due time, it will be done. Once recognized and undertaken by the scientific world, it will clearly demonstrate the validity of the New Universe Model.

Deceleration proof: All of the rotating and orbiting in the universe is angular momentum that was extracted from energy that was originally linear momentum. There are no other angular momentum sources. This, by logic, is direct evidence (proof?) that the universe's objects can only be decelerating.

4. Production of the Universe; Continuing matter production and growth

*(Matter: includes **all** energy **and** mass)*

Nuclides (Elements)

There are 110 named + 4 elements known to exist [15, 29, 30], and all are listed in the Periodic Table of Elements, which is printed in almost every basic chemistry book (also in the book "Chart of the Nuclides" [15]). Chemical elements are substances containing a specific number of protons in the nucleus of each atom. The nuclei of all of the elements also contain one or more neutrons. (Except the basic Hydrogen nucleus, it contains only one proton) All atoms/ nuclides are elements which contain a specific number of protons, but each can have different numbers of neutrons; and these nuclides are called isotopes of the base element.) Over 3600 known nuclides are listed in the Chart of the Nuclides [15].

Most elements heavier than lithium but lighter than iron, were produced by fusion of protons and neutrons in the high temperature and high pressure 'furnace' core of first, second, and third generation stars. Heavier than Iron elements were produced in and during the high pressure explosion blast in supernovae.
(Fusion: A nuclear reaction in which nuclei combine to form more massive nuclei, with the simultaneous release or absorption of energy)

Star core manufactured elements are initially dispersed by supernovae explosions of their progenitor star. Fusions in stars' cores and the explosive force within supernovae produce and disperse the heavier than iron elements. We find these in our third (or fourth) generation sun and its satellites. Nuclides that absorb more energy than they release in the fusion process were made exclusively from the high blast pressure and temperature within and during the supernovae of their progenitor star.

It is generally accepted [15],[40] that the population of nuclides in the universe, before processing within the fusion furnaces of stars, was approximately 75% Hydrogen, 24% Helium, and < 1% Lithium, with only a few slightly heavier nuclides.

It has been theorized by a few physicists that dark matter, often referred to as Weakly Interacting Mass Particles, (or simply WIMPS,) are exclusively made of multiple neutron nuclides, which do not contain any protons. They were named nucleonos in a paper written in the 1990's by Ken C Freeman of the Mt Stromlo observatory in Australia. (Dark matter has been proved to exist, … not to be confused with the mythical dark energy)

(An individual neutron has a half life of only 10.25 minutes when it decays into a proton and an electron, plus an energy wave [15]. It is presumed that in a multiple neutron nuclide the neutrons are stabilized and have an unknown but very long half life, as they do when they are included in most other nuclides, such as helium.)

To produce an element heavier than hydrogen, proton and neutron particles must be fused, which requires a force be applied, either by velocity enhanced gravity attraction pressure; or by compression momentum as occurs when two particles collide. Adequate momentum for collision fusing requires temperatures of ~100 million degrees Kelvin. Temperature in fluids, gasses or liquids, is a measure of the mean velocity of the particles. Fusion force by velocity enhanced gravity requires ~parallel velocities of 99.9999999 % of c, (speed of light). (See "Deflagration Progression … with Laws of Physics" along top of Figure 4, pg 15)

Including the multiple neutron nuclides (^{2,3,4}n) that make up the dark matter 'WIMPS', the universe's mass population was originally; ^{2}n = 87.5%, H = 9.38%, He = 3%, and only .22% Li and other nuclides.

Origin of matter in the universe:

Theories for the origin of the universe are the <u>BB</u> and this new model (<u>NUT</u>). Both presume that all physical matter, (sub-elementary particles to the largest objects including stars and galaxies), precipitate and condense from energy in the form of electromagnetic waves.

The origin of the energy for the <u>BB theory</u> is presumed to have come from 'nothing', instantaneously, at a single dimensionless point. (The postulated <u>BB concept inconsistent with Laws of Physics</u> and of course cannot be proved)

The origin of energy for the <u>NUT concept</u>, and now the New Model of the Universe, contends that it originates from annihilating positroniums, which by logic, are thought to occupy all space before the universe started forming. 'Positronium' is the name of a pair of mutually orbiting leptons, an electron and a positron, held together by opposite electrical charges and their mutual gravities: their separation is maintained by their rotational centrifugal forces. Positrons were first postulated in 1928 by Paul Dirac [41] and proved and observed experimentally in 1932. Positron and electron annihilations have been observed in laboratories and nature.

Inside the universe, individual positroniums are unstable with a half life of less than a second, but outside the universe where there are no random waves, positronium components are stabilized by uniform separations as they revolve about each other in synchronization with adjacent positroniums in a continuous array. Such separation geometry in chemistry is called a uniform hexahedron lattice crystal. Equal separations at 6.25 cm in all three dimensions results in precisely the same mass density as found inside the universe, including dark matter WIMPS.

For the Laws Of Physics (LOP) NUT model, nothing is created or destroyed. Therefore matter (electrons) and antimatter (positrons) are presumed to be the basic primordial matter material. Thus, the <u>NUT concept is totally consistent with the Laws of Physics</u>, with nothing destroyed or created, as stipulated by the first law of thermodynamics. Therefore, the <u>LOP NUT model is feasible and plausible.</u>

The BB model *(Big Bang theory)*:

Steven Weinberg, a credible nuclear physicist, also developed the 'matter process model' for the BB theory. Weinberg accepted the BB hypothetical presumption of everything ... all initially as energy, came from an un-definable nothing. His analysis starts after the initial moment, and accepts hypothetical assumptions about temperature and expansion velocities.

In particle accelerator laboratories, particles are collided at very high velocities to fragment the accelerated and target particles into their subatomic components. Weinberg's analyses uses the experimentally proven processes in reverse to calculate and estimate how an almost infinite energy package could transform energy through various steps into very small particles and at high temperatures fuse into the matter (energy and mass objects) that occupy the universe today. His analyses start a short time increment after the presumption of the BB theory's 'initial moment', which was the eruption of a humongous amount of energy. A<u>ll</u> the mass and energy that exists, is said to have erupted out of nothing, and <u>all</u> of it is claimed by the BB theory to have come from one infinitesimal point at one instant!

The assumed hot dense beginning was postulated in the early 1900's by astrophysicists A. Friedman and Georges Lemaitre. The conditions (temperature, density) for the 'magical big bang' start, for lack of a definable term, is sometimes referred to by a vague un-definable catch-all term; '<u>quantum anomaly</u>'! 'The BB' analyses cannot explain the initial moment or the continuing acceleration of all space and objects.

BB theory advocates assume matter (both mass and energy) transferred from radiation energy into mass particles while the matter was dispersing, spherically radiating in all directions, into <u>all</u> the matter that exists today. The transformed matter had to cool by expansion, transform into nuclides and elements and then coalesce into larger massive objects (stars) before the particles' separations were too great. The window for this occurrence was small and it appears that if WIMPs are multiple neutron nuclides as speculated, the window was too small for the production of all the universe's observed matter (by a factor of about 8).

Weinberg's model required more time than the assumed big bang-to-now period could accommodate. The BB was about to be scrapped, but at that time there was no other option on the table and along came a bright young physicist named Alan Guth who rescued the BB. He theorized that if a portion of the expansion occurred at a rate faster than the speed of light, the BB concept processes could work. He, and the other BB theory believers, refers to their rapid hypothetical expansion period as 'inflation'. Somehow!, after a time, the rapid inflation-expansion rates at and above the speed of light are disregarded, and then they start explaining transmutation of mass and energy of particles at simply high temperature. Somehow the whole BB expansion process slowed and restarted accelerating again. After ~13,700+ million years it is (mis)interpreted and presumed to be expanding at the red-shift vs distance Hubble number. It is erroneously assumed that the H# is an acceleration rate for all matter; all without any, or from a hypothetical motivating force.

In summary, <u>the BB theory model violates pertinent Laws of Physics</u>;
Anything outside the L o P is not real; i.e., it is mythical and imaginary.
1. Matter (both mass and energy) cannot be created or destroyed.
2. Inflation assumed expansion rates higher than the speed of light.
3. Expansion assumes all objects and space are continuously increasing their velocity and accelerating without any force.
4. Origin and escape of <u>all</u> matter from a small point much smaller than the Schwarzschild radius of even our little Earth (.9 cm).
5. As of 2009, time lapse z data indicates accelerating expansion is not real.

<u>Graphical 'History of the Universe' as presumed and postulated by advocates of the Big Bang theory (BB)!</u>

A graphical drawing of the BB theory as sanctioned by the Particle Data Group of Lawrence Berkeley National Laboratory (LBNL) is replicated: Big Bang theory believers call it the 'history of the universe'. It presumes energy is transformed after the BB theory's mythical 'quantum anomaly' event into other forms of matter, both mass objects and various forms of energy.

Process descriptions which Weinberg calls 'frame 4 through 7' in his book "The First Three Minutes" [16], are not only feasible, but are similar to what is occurring in Stage II of the LOP NUT deflagration wave processes. In the BB theory, rapidly diminishing fusion rates due to theorized expansion allow only a very short time window for a few 'light' element nuclides to be produced.

Weinberg's predictions correlate with the observed element populations (or visa versa), which some astrophysicists believe coalesced into the first generation stars. That is where heavier nuclides were fused and later dispersed by supernovae, subsequently, again coalescing into formation of descendent star generations. Most of the elements heavier than Helium were produced in the fusion furnace core of first and second generation stars. Fusions in star cores and pressure fusion during supernovae of second and third generation stars were required to produce the elements we find in our third (or fourth) generation sun and its satellites.

The "History of the Universe" [31] chart artistically illustrates Steven Weinberg's concept of matter transmutation <u>after</u> the BB theory's postulated and presumed infinite energy source converted into <u>all</u> of the mass objects and energy in the universe of today. Weinberg's concept of energy-to-mass better fits gravity enhanced fusions than particle collision fusions from super-extreme temperature (high velocity particle collisions).

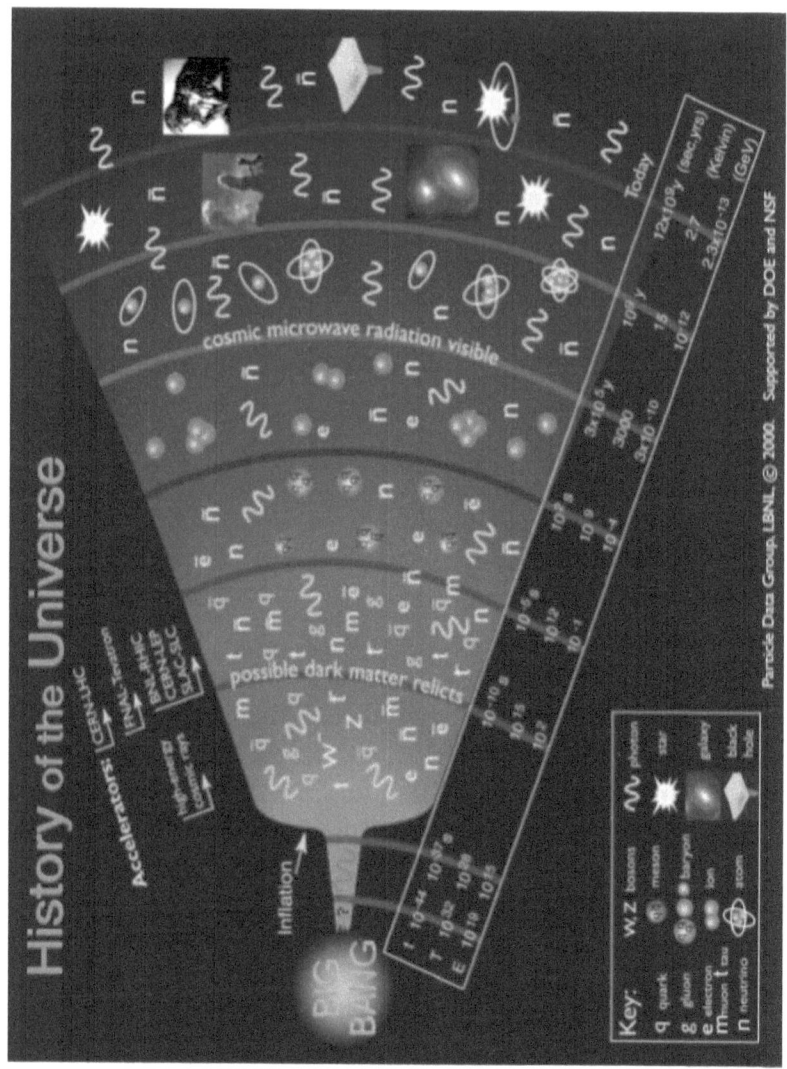

Figure #13

Chart summary: After the ' infinite' temperature BB theory matter cooled to the hypothesized temperature of > 10 ^{32} Kelvin, it cools to ~ 10 ^{24} K, energy then transmutes into sub-elementary particles as it cools to about 10 ^{12} K. These sub-elementary particles fuse into leptons as it cools to 10 ^{9} K and then as the matter mix cools towards ~>3x10 ^{3} K, leptons fuse into hadrons. After matter temperatures further cool to less than 10^3, clumping begins for the formation of stars, etc.

Laws of Physics (LOP) Model of the Universe:

Because of my strong belief that the universe developed in a scientifically credible manner, it has been my desire to identify the physics compliant processes from which it came into being. The chronological feasible proven processes only needed to be identified. To be scientifically credible, all processes necessarily must be consistent with the Laws of Physics and not assume any hypothetical, non-provable or non- existing phenomena. This LOP feasible and scientifically credible model of the universe is deducted from "New Universe Theory with Laws of Physics" [1].

Now, with observations and verification, the universe continues to develop and grow, not expand with acceleration. It grows and develops, from the processes in the Stage II deflagration wave. The wave progresses through Stage I primordial matter; the annihilation energy transforms into the energy and physical matter of the Stage III universe, where we and all observable objects reside.

Electromagnetic radiation is produced from the positronium annihilations of positrons and electrons, just as demonstrated in particle laboratories. (511,000 electron volts of energy are transmuted from each particle). Primordial matter is described and explained as positroniums which are stabilized by their uniform spacing, orientation, and synchronized revolutions. Feasibility of this stable primordial matter is scientifically described in the NUT Book [1].

Known observable universe original element nuclide populations (~75% H & ~24% He) [35] are precipitated and continue to precipitate directly from outward flowing gamma wave electromagnetic radiation within the deflagration front which completely surrounds the universe within. Similar transformations via $E = m\,c^2$ are being demonstrated and researched in the BEPC Laboratory (Bejing Electron Positron Collider [19]). The BEPC is a billion + dollar facility, built and operated in Bejing, China.

The LOP New Universe Model transmutation processes are similar to the Weinberg BB theory analyses as pictorially shown in the DOE & NSF sponsored LBNL artistic 'History (?!) of the Universe' illustration (Figure 13). However, more plausibly, the photon precipitating subatomic elementary particles fuse into nuclides from the force of **Newton gravity laws and Einstein relativity**. At high velocities, 'velocity enhanced mutual gravities provide the fusion force', not collisions of high temperature (momentum) particles. In the LOP New Universe Model concept, photons precipitate into sub-elementary particles, fuse into, and precipitate as Hydrogen, multiple neutron nuclides, Helium, Lithium and a few other nuclides. (As found in the universe)

It has been demonstrated by angular momentum analyses of individual galaxies; there exists about eight times more mass in the universe than is directly observable today. It is thought to be in the form of nutralinos (multiple neutron nuclides) in and around galaxies. This dark matter is theorized by a few physicists to be made of multiple neutron nuclides, which do not contain any protons, and are referred to as nucleonos. As stated earlier, an individual neutron has a half life of only 10.25 minutes and then decays into a proton and an electron. Apparently however, in a multiple neutron nuclide the neutrons are stabilized and have a much longer half life, as they are known to have in most isotope nuclides. To produce an element heavier than hydrogen, proton and neutron particles must be fused together, which requires a force be applied, either by enhanced gravity force, pressure, or momentum exchange, as occurs when particles frequently collide in all fluids.

Fusion productions soon diminish behind the deflagration wave due to reductions of velocity enhanced gravity force because linear velocity is transformed into angular momentums and fusion heat. These decelerations are similar to what occurs in all other 'Fluid flow' decelerations, … from vortexing and mixing.

The Deflagration Wave front propagates outward, being led by additional chain reaction annihilations of primordial matter positron-electron pairs. The annihilation photons propagate in all directions and are uniformly spherically spread (throughout 4 Pi steradians). As more annihilations occur at the forefront of the deflagration wave, only those photons that are radiated in the forward ~2 Pi steradian hemisphere increase in intensity/density and accumulate into concentrations such that baryon particles are transmuted. The density of photons that radiate and flow in the opposite-rearward direction of deflagration do not intensify and propagations are in the direction towards the universe; these are left to become part of 'background radiation'. We can only observe photons that have traveled to us from the front which is traveling outward, also at light speed away from us. Therefore the front is currently, and always will be, twice its potentially observable distance from us.

Primordial Matter Concept

Recognizing that red-shift is an accurate measure of velocity, and not a measure of acceleration, it becomes clear that expansion, proclaimed by Hubble 'as obvious' in the 1920's, was strictly an illusion which is inconsistent with scientific thinking, since it violates known and proven laws of physics. The increase of red shift between farther away individual galaxies with distance (not time) was first observed and demonstrated by V M Slipher in 1912 at Lowell Observatory.

It is now explained the farther away galaxies are decelerating in an outward flow direction from a farther out faster moving source, the deflagration wave. The universe is growing (more mass adding onto the periphery), not by space expanding inside the universe.

Converter of Primordial to Universe Matter

Physics of Stages and Sub-stages are explained in reference book "New Universe Theory with Laws of Physics [1]"; "Which is it?"

Figure #14

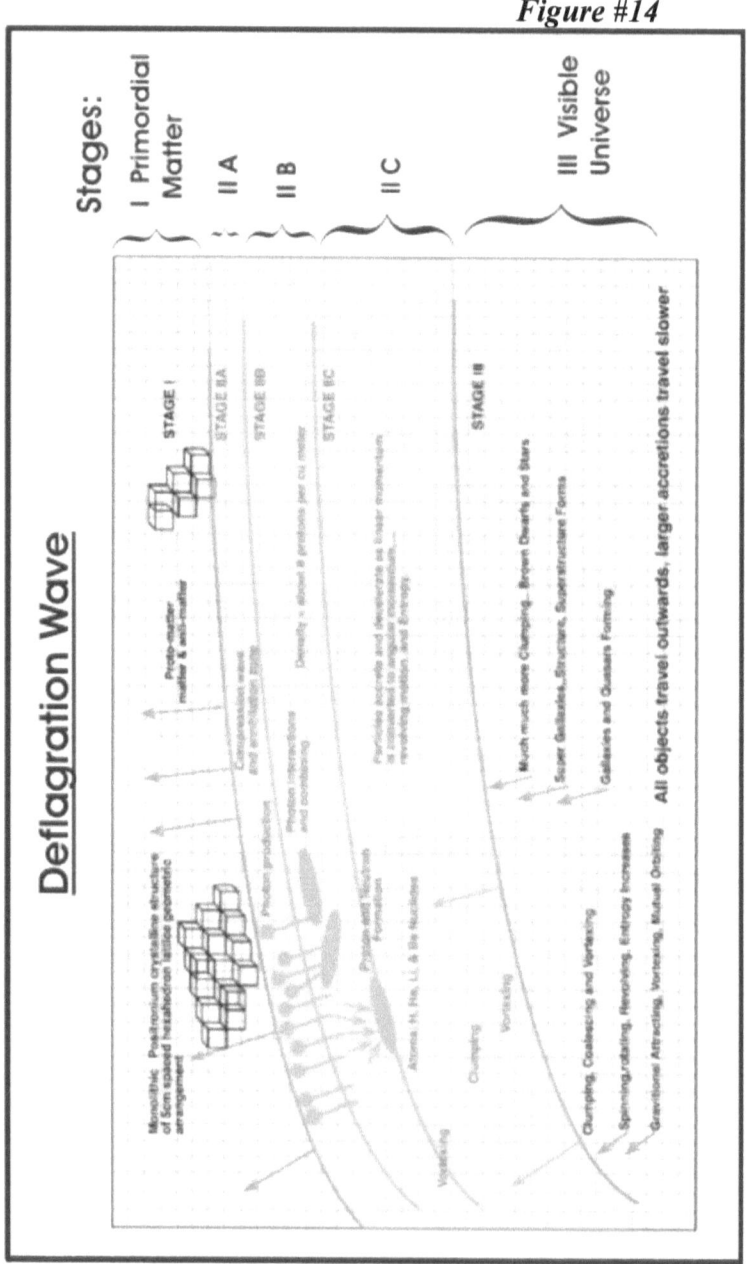

Deflagration Wave Processes

Function processes in the deflagration wave are; Newton's <u>mutual mass gravitational attraction force</u> and Einstein's <u>mass-velocity relativity</u>. Both are physical laws pertinent to the origin and development of the universe and its contents. (See boxes in graph)

Figure #15

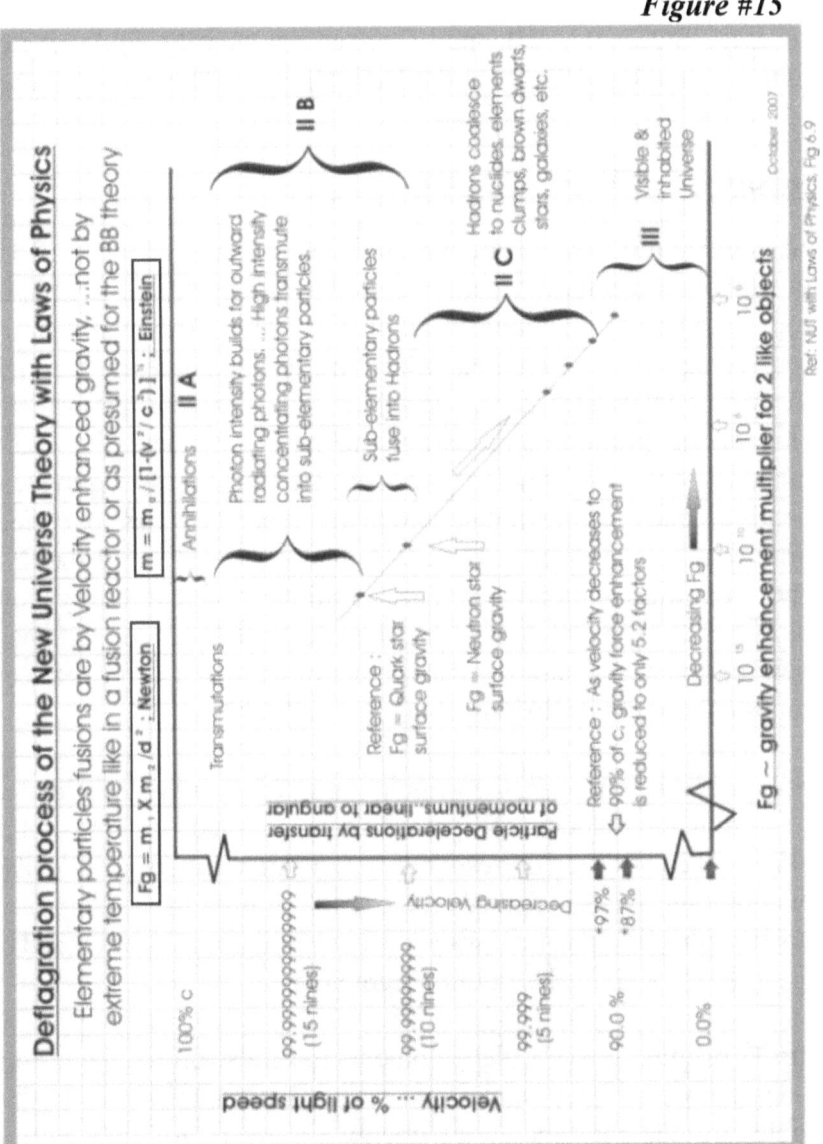

To graphically illustrate (Figure 15) the two most pertinent phenomena in the deflagration wave process, (velocity and gravity force) it was necessary to use multi-log vs multi-log scales. (for easier comprehension, right-to-left) The divisions on both the ordinate axis and abscissa axis are multiple log-log cycles. Multi-log-log graphs illustrate the vast values that achieve the fusion force processes. Stages IIA, IIB, and IIC are comprehensibly and verbally described in the reference book "New Universe Theory with Laws of Physics".

Deflagration Wave Thickness

The deflagration wave thickness includes the distance from gyrostablized primordial matter at zero velocity to the first light propagating stars, traveling at about 97% of c. Annihilations are progressing in all directions away from the universe, through the primordial matter at the speed of light. Photons perturb the synchronous primordial positroniums and perpetuate the chain reaction front. The photons concentrate into mass equivalent quantities, and then precipitate into (sub) elementary mass particles. Particles then coalesce by velocity enhanced gravity until the first energy emitting stars appear. They continue deceleration by flow mixing; that is, by collisions and near collisions to convert their linear momentums into angular momentums in stars and between groups of stars; by collisions, rotations, and orbiting.

From inside the universe, the velocity range of the wave is estimated to be from the velocity of light 'c' down to about ~ .97% of the velocity of light which corresponds to red shift of about $Z = 27$. The deflagration wave is indicated on the graph (Figure 14) as Stage IIA, plus IIB, plus IIC. The total wave distance / thickness is estimated to be about two and one half billion light years.

During original concept development, the most difficult question to answer was; 'from what could primordial matter be made that would provide the amount of energy necessary for transmuting and processing into the matter in the universe?' The only non-mythical source for that gargantuan amount of energy can only be; 'It came from <u>many</u> matter / antimatter annihilations. This was the only known source of high intensity energy packets for sufficient subsequent energy for mass transformations into the universe that we observe'.

Primordial Matter

Primordial matter that could feasibly produce the universe could only be made up of several possible combinations of matter and anti-matter, stabilized in a crystalline structure arrangement and separated such that the matter density is equal on both sides of a transcending deflagration wave. The laboratory proven, simplest and most straight forward are positroniums. The equally spaced hexahedron arrangement could provide the dynamic stability, and as such, they are not subject to the short half-life we find for positroniums here in the universe.

In engineering and nature, the simplest geometry usually provides the best solution. ([1] Figures 6.4 & 6.5). Research on annihilations and energy transmutation to mass is currently being conducted at the BEPC (Beijing Positron Electron Collider) research laboratory. Their work strongly supports the LOP New Universe model concept. That 'leading edge' laboratory has been in operation for several decades by the Chinese government. A major upgrade (~ 1 billion dollars) has recently been funded and is in progress. It will increase the possible number of concurrent electron-positron annihilations. The upgraded BEPC will make possible increased intensities of gamma ray photons which are expected to transmute into larger than previously produced elementary subatomic particles. It is expected they ultimately may be capable of demonstrating quark production, and hopefully, but not likely, produce protons and neutrons.

The BEPC annihilations and transmutations are identical to those described for primordial matter and transmutations at the leading regions of the deflagration wave! (Stage II, Zones A and B, shown in reference [1] Figure 6.14). I was unaware of the BEPC at the time of the LOP NUT development, but the BEPC upgrade operational results will go a long way toward supporting, confirming and validating the LOP NUT origin of the universe concept.

Origin of primordial matter and the initiation of the first annihilation to start the deflagration wave are neither known nor even postulated.

Subsequent from initial annihilation and origin of primordial matter, all processes are proven to be scientifically plausible and are clearly feasible within the known and proven Laws of Physics. Development of a theory that precedes primordial matter and events before the deflagration wave; like some of my college text books stated, proof is left for the student! (The ahead of its time; 'string theory' might contribute to future understanding?)

In summary, the LOP <u>New Universe Model complies with all laws of physics</u>; (It does not include hypothetical 'dark energy').

1. A large volume, if not all, of space outside the universe, is full of an array of hexahedron arranged positroniums rotating in sync and equally spaced at 6.25 cm. in all three dimensions.

2. One single positronium became out-of-sync; ... the ensuing annihilation triggered a dynamic imbalance and annihilations chain reactions spherically radiating deflagration wave that converts positroniums into energy, then into mass particles. (Figure 15)

3. Velocity enhanced gravity coalesces and fuses particles. Vortexing produces larger / compound particles and <u>all</u> the matter (mass objects, energy) of the universe.

Former New Universe Theory [1] is now referred to as the <u>Laws of Physics Model of the Universe</u> since it is now <u>proven</u> valid.

---

5. Cosmic Background Radiation

The only reason background radiation is discussed here is that many of the BB supporters claim it verifies the BB. Nonsense; Background radiation proves that the matter in the universe precipitated (inefficiently) from energy which was initially in the form of gamma rays. This does not verify the source of the energy for transmutation.

Cosmic Background is claimed to support the BB Theory, but COBE and WMAP cosmic Background survey findings better support the LOP NUT Model. Spectral distribution of radiant heat energy from any black body was predicted by Max Plank in 1901 and now is considered as law for energy radiation from black bodies. The 'black body' name applies to all objects and was so named to differentiate between a body's radiation energy and reflected energy. (Even the brightest of stars are black bodies by the definition). Plank curve amplitude defines the temperature.

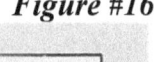

Figure #16

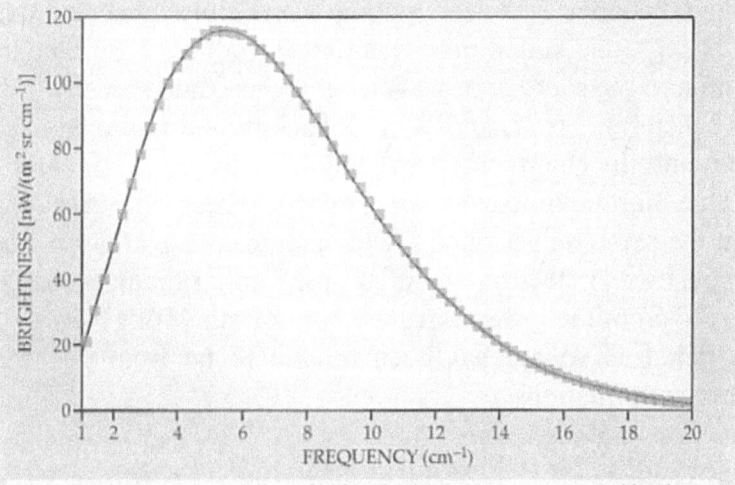

Figure 1. The cosmic microwave background spectrum as measured by the *Cosmic Background Explorer*'s FIRAS spectrometer and presented by John Mather in January 1990, just eight weeks after *COBE*'s launch.[1] Boxes represent conservative estimates of measurement uncertainties. The data show no discernable departure from a perfect blackbody spectrum (the curve) with a best-fit temperature of 2.735 ± 0.06 K.

When the data measured by the COBE and WMAP projects are plotted on Plank's black body spectrum curve, it fits. But of course, the distant universe is by definition a black body. The thermal properties in our region of space in the universe can be called 'left over from deflagration precipitations and nucleosynthesis, and are now quantitatively known, thanks to WMAP and COBE projects. The LOP NUT deflagration wave annihilations of Stage II A and the subsequent first generation brown dwarfs and stars decelerating out of Stage II B are sources of the initial Plank radiation, which is more plausible than the hypothetical single point origin theory. The original BB temperatures are construed from the COBE results using Boyles Law for gases (PV = RT) and assumed expansion of the universe which is now proven to <u>not</u> be part of the universe development process.

The deflagration front process positronium annihilations result in gamma wave pairs of 511 keV gamma waves which are the matter in energy packets that precipitate and transmute into the particles that coalesce and transform into all of the physical matter throughout the universe. All of the annihilation gamma waves are not consumed by transmutation; some simply become part of the 'background radiation'. Gamma waves are electromagnetic radiation in the shortest of wavelengths, less than ~ one Angstrom (10^{-10} meters). COBE and WMAP background radiation surveys have found the gamma ray wavelengths between .03 nm and .003 nm. One nm (nanometer) is 10^{-9} meters. These nm wave lengths are in the same range found for the gamma waves that propagate from positron – electron annihilations. Similar radiation has been observed from the glow above the core of our Milky Way galaxy. That radiation source has been proved to be from positron & electron annihilations.

Plank Black Body

The black body radiation curves in the following chart were calculated using Plank's Law equations for five black body temperatures at 3500 °K through 5500 °K. (Spectral distribution of black body radiation illustrations are from Wikipedia, the public domain on-line encyclopedia)

Figure #17

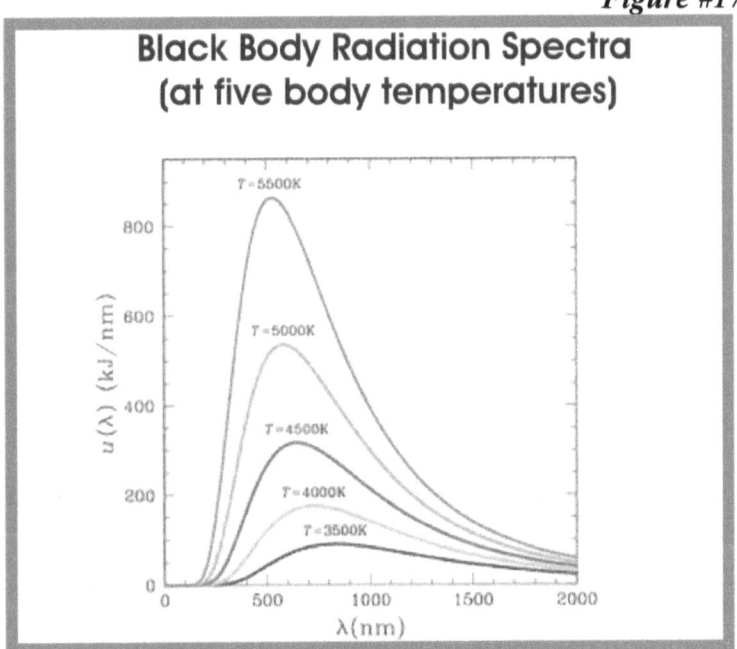

The radiation curve for a black body at 2.7 °K is similar except for amplitude. The cosmic radiation data was found to correlate with that curve very closely. Those results show that the cosmic background is a typical black body radiation source, consistent with and independent of any origin of the universe concept. The radiation energy level is a function of the source temperature and distance to origin of the source radiation. Logically, this has to be where the trailing edge of the deflagration wave was, at the time the currently measured radiation from the wave's rearward radiation as it progressed through and beyond our region of the universe. Large scale anisotropy, more definitively called asymmetry, is to be expected because we are not at the center.

Smaller scale anisotropy is because much of the radiation was randomly transmuted into mass particles as photon intensities become adequate for precipitating into particles, and culminate into brown dwarfs. Ultimately, these accrete more matter and become stars, galaxies and quasars, and other electromagnetic energy radiating objects. The data as presented is consistent with what is expected from the NUT. Survey results better support the LOP NUT than the BB.

In the LOP New Model of the Universe, as described in the NUT, forward radiating photon concentrations transform and coalesce into elementary particles and then into protons, neutrons, nuclides, etc. This occurs within and starting at the forefront of the deflagration wave which completely surrounds the universe.

Potentially, it can only be observed at ½ its distance away, because it is traveling near the speed of light, the same speed that the radiation from it travels to us. It is currently thought to be about 30+ billion light years away, but only observable up to ½ that. Current theory is that the background radiation source is 'left over' from the origin. The LOP NUT is consistent with that hypothesis.

Deflagration front chain reaction annihilations occur from stationary* positroniums, therefore we should see very little red-shift from their photons; only that due to our region's recession. Rearward (from the deflagration wave) radiating photons will be red-shifted only to the extent of the 'impact' perturbation of the positrons and electrons that causes the matter / antimatter contacts. Only photons radiating forward will increase in quantity and concentrate to produce sub-elementary particles, combine, and produce the objects of the universe. The annihilation photons that radiate in other directions will either be absorbed by mass objects or become part of 'background radiation'.
(* *stationary relative to the center of the universe and the external-universe primordial structure*)

6. ASTRONOMICAL KNOWLEDGE: Now

Where have we been and how/what have we been thinking for the last three thousand years? Scientific thinkers have contributed to the progress of all true / real knowledge in astronomy, with realism over myth; but only with great sacrifice. Many of our beliefs continue to be based on not provable hypotheses. As long as science accepts some ideas, such as 'dark energy', the door remains wide open that invites other mythical postulations.

FIRST MILLENNIUM: 000 ~1000 *(Flat Earth Millennium)*

Hypatia: (370-415)... Hypatia of Alexandria, possibly the smartest person that ever lived! She was a speculative and theoretical philosopher, mathematician, and astronomer who lived and taught in Alexandria, Egypt. She was curator of the greatest library of the time where she accumulated much information about the night sky, including knowledge about the Andromeda galaxy, speculating that it was another 'universe'. Her superior knowledge embarrassed the Archbishop. A riot was incited and unfortunately she was horrifically murdered in 415 AD, and her library was burned. For the next 800 years, the dark ages ensued and little progress of knowledge was made during that era.

SECOND MILLENNIUM: ~ 1000 to ~ 2000

Copernicus: (1473-1543) ... Nicolaus Copernicus was the Polish astronomer who advanced the theory that Earth and the other planets revolve around the Sun, disrupting the Ptolemaic system of astronomy. Ptolemy was an Alexandrian astronomer, geographer, and mathematician, who based his ideas of astronomy on the belief that all heavenly bodies revolve around the earth, as implied and explained in the first 17 verses of Genesis. Since the days when those words were written, with the aid of modern technology and with scientific thinking, mankind has developed more knowledgeable explanations.

Galileo: (1564-1642)… Galileo Galilei was a great Italian astronomer and physicist who was first to use a telescope to study the stars (1610). He was an outspoken advocate of Copernicus's theory that the sun forms the center of the Solar system, which he proved with use of his telescope. This discovery disproved the Bible. Sadly, he was persecuted and imprisoned for the remainder of his life [32] and he was prohibited from publishing his findings. After his death, his notes and records became available via his nun sister.

Slipher: (1875-1969)… Vesto Melvin Slipher was chief astronomer at the Percival Lowell (1855-1916) Observatory near Flagstaff, Arizona where he was the first to use a spectroscope coupled with a telescope to study star spectra and thereby determine the elemental constituents of stars. He discovered galactic Doppler red-shift in 1912, and further determined that the farther away a galaxy, the faster it is receding from our location in the universe. His findings were interpreted by others to theorize universe expansion.

Hubble: (1889-1953)… Edwin Hubble claimed the 'universe is expanding', based on the Slipher discovered and measured red-shift data. Hubble and Milton Humason used new measurements of distance by using 'Cepheid Variable' stars as standard brightness candles in their host galaxies. This was over ten years after Slipher's discovery. In 1923, Edwin Hubble and Milton Humason confirmed Slipher's 1912 finding that the more the red-shift the farther away the galaxy, proving higher galactic recessional velocities with larger distances. He proceeded to name the apparent rate of expansion as the 'Hubble Number'. That misled to the non-provable and implausible idea that all matter in the universe came from a single point. That hypothetical postulation is inconsistent with laws of physics, and was named by Sir Fred Hoyle with a derogatorily intended reference; 'BB theory'! But the 'BB' name stuck.

THIRD MILLENNIUM: ~ 2000 – 3000

The Laws of Physics (LOP) NUT millennium

In the 21st century (the start of the third millennium), many astronomical facts are being discovered, but very little of the new knowledge that exists is yet being accepted. Beyond the first decade of the third millennium, the LOP NUT scientifically explains the universe's origin and ongoing cosmology. Many mythical ideas e.g., dark energy, continue, at least at the beginning of this third millennium to get as much study consideration as do Laws of Physics founded ideas. We now know that the universe is not expanding with acceleration, so Boyle's Law for gases does not apply to the universe as proven by simple studies referenced and presented in this document. The background temperature was construed as proof of the BB origin. Now we know better.

In today's society, fortunately, we are allowed to consider new concepts without the fear of Hypatia's demise. I am hopeful that the LOP NUT "Model of the Universe" will advance scientific thinking without myth among all civilizations. Such non-myth progress might let the Human Race come together and exist to its potential. (see Time line bar chart, Figure 3)

Future:

Understanding the Universe and its origin have been a major matter of speculation and theorizing ever since mankind started pondering the night sky. Astronomy has been plagued with mystery and non-scientific myth explanations of the unknown. With the advent of scientific thinking, knowledge slowly began to overcome myth early in history. We are now in the third millennium, and it is time for us to proceed ... with scientific thinking: ... with only proven facts, and with the tested and never disproved Laws of Physics.

7. Ongoing studies ... New Model

7A. <u>Growth</u>: (continuing radial addition of space and matter in three dimensions)

Obtain more far distant Sn1a 'Z' data, also review distance basis; (i.e., latest intrinsic brightness considering Nickel 56 influence)

When we recognize we are not at the 'center' and neither is 'everything' (as presumed by the BB), we ask, where did it start and where is the center now?

Even though the new model shows the universe is made up of mostly mass objects which are matter precipitated from the deflagration wave as it passes through primordial matter, the wave started from somewhere. One approach to finding the answer was to start with the Hubble lines. By measuring the min and max, we should be able to calculate the distance we are from the center, and pin point the origin as well as our position. Allowance must be made recognizing we can only measure the radial component of an object's recession as observed from our site, and objects should be radiating from the origin, even though they are receding from us as well.

Since we are not at the center, there should be different H# lines in different directions. Although the Hubble lines are different in different directions, they would not appear to be separated in the distance, just as railroad tracks appear to merge in the distance. Originally the H# lines were assumed to be linear, but when extrapolated to the speed of light the distance between the minimum accepted H# and the max H# revealed a distance that was too small to yield the minimum universe diameter. H# lines were then reconciled to provide the then thought to be diameter of the universe. In year 2006 Data was acquired by a study team (actually they were looking for the 'mythical dark energy') that showed the curvature of the H lines consistent with the 'reconciled lines'.

After recognizing that the H lines are gradient lines across lines of deceleration, it was recognized the H# lines would be asymptotic at the trailing edge of the deflagration wave, which is propagating near the speed of light. Another study by the "dark energy' study team produced additional (published in 2009) H# line deflection which corresponds to the H# gradient line as predicted by this model of the universe concept. This further substantiates the New Universe Model.

The following charts illustrate the universe in development following the precipitation of astronomically observable matter and objects from the deflagration wave, shown as propagating across the top of the chart.

The first graphing of the NUT ... Max and min Hubble lines

Figure #18

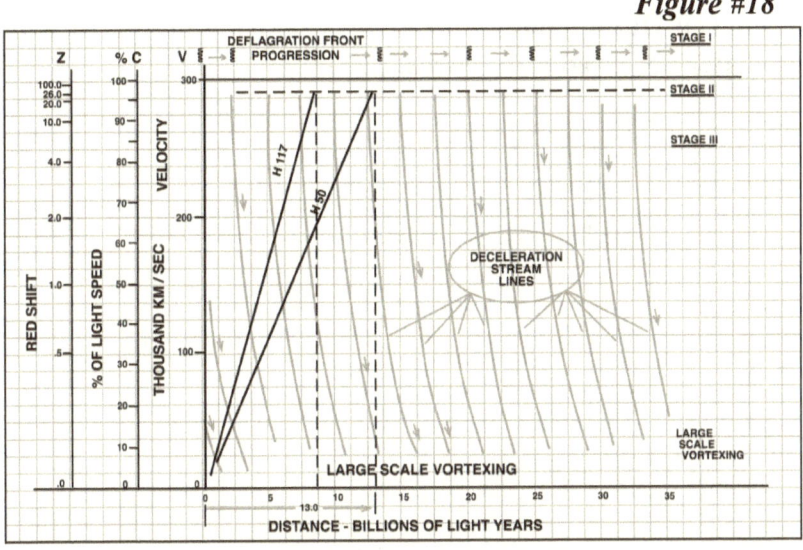

Figure 18 shows the Hubble lines as originally drawn to show the H 50 and the H 117 lines of 'expansion' as assumed by the BB 'expansion theorists'. In this view those two lines are overlaying the travel lines of galaxies as they decelerate from the deflagration wave.

Figure 19 is a modification of the Figure 18 chart to illustrate how Hubble lines are not straight; they are curved gradient lines. The H# lines are actually lines through different galaxies that lose velocity with time and distance due to linear momentum being converted into angular momentum. H# lines are deceleration gradient lines, not galaxy acceleration lines as presumed by the BB concept.

Lines of estimated deceleration are superimposed on both charts, and individual galaxies are illustrated by dots on the reconciled Hubble line for H# 50. The reconciled lines coincide with data as expected to be measured in the future with new technology telescopes. The arrows have been added to the original 2004 Reference [1], Fig.10.1 chart. The large shaded arrows show study results published in 2006 by a 'dark energy' research team. [6]

Reconciled Hubble lines with arrows showing predicted which were later observed differences.

Figure 19

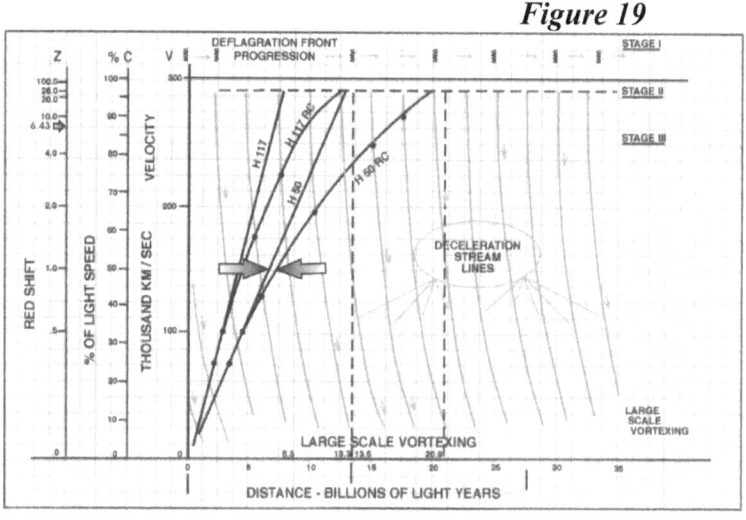

'Dark energy' study teams are not only verifying the New Model of the Universe, they are calibrating the gradients, which will eventually lead to defining the universe's flow coefficients. The deceleration stream lines will then be definable. Figure 18 and 19 deceleration lines are of course only estimated at this time.

Figure 20 shows what is currently expected to illustrate the universe development and growth. The 'fat' arrows with #5 are observations (published in 2008) by the 'dark energy' research team [7]. The fat arrow with #2 is my estimate of the full H line.

Arrow descriptions follow the chart.

Figure #20

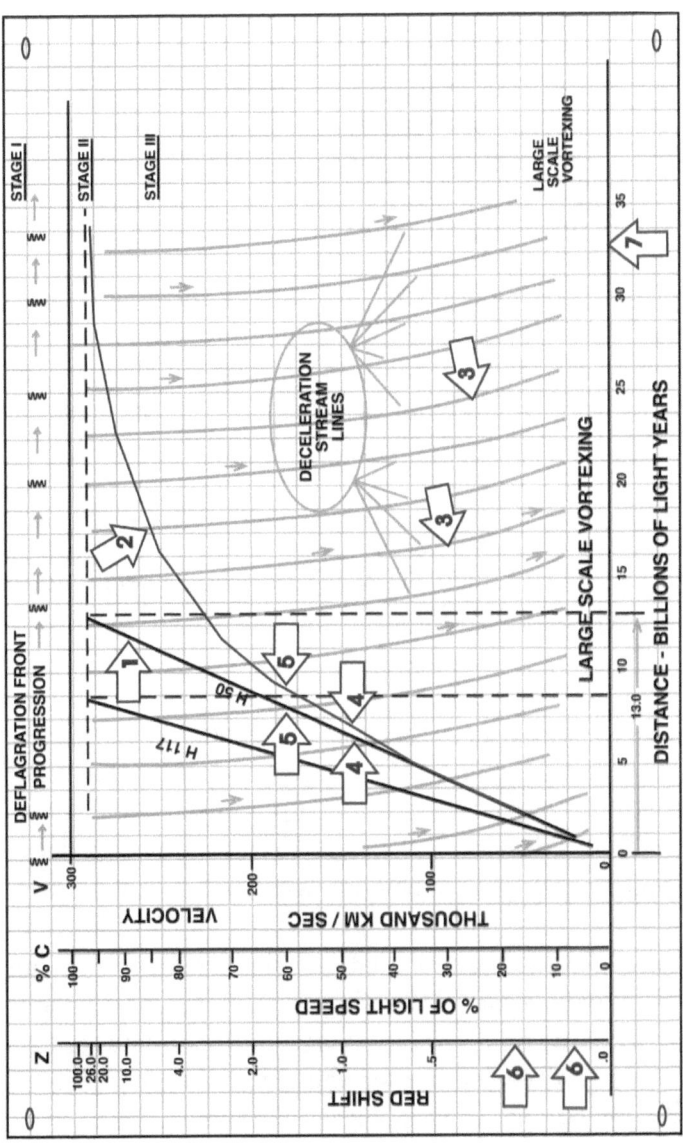

Measuring the H Gradient Line to the edge of the universe (the deflagration wave) will be exciting and fun for astronomers.

Arrow #:	Description:
1	Original typical Hubble 'expansion' line
2	Reconciled H Gradient line, estimated
3	Estimated Deceleration stream lines
4	A Reiss Team #1 [6] 2006 & 2007 data
5	A Reiss Team #2 [7] 2008 data
6	Abell catalog Z range of 4076 clusters
7	Current universe size correlates with age (Fig. 3)

Very worthwhile achievement!

'Dark energy' research teams are not finding dark energy, but more importantly, their continued studies are defining the universe growth characteristic. Extending the reconciled Hubble line to the deflagration wave requires even farther distant data, like is being obtained by the "dark energy" team. (Figure 20, Arrow 2)

7B. <u>Expansion</u> (Implies accelerating growth)

Recommend: Research ACO & SR time lapse verifications
(Suggested in Topic 3; analyses in Appendix 2)

Changing velocity cannot be measured directly, but the red-shift tool can be used to <u>disprove</u> increasing separation velocity between us and any specific galaxy cluster, as claimed by the BB theory. Proof simply requires two or more <u>time lapse</u> separated red-shift measurements, repeated for the same galactic cluster. Selection of a cluster's alpha galaxy must be done with care.

1920's and 1930's astronomical red-shift data proves only that farther away galaxies are separating faster than closer galaxies. The only way to answer the velocity increasing with time (acceleration) as opposed to no acceleration or velocity decreasing (deceleration) is simply by comparison of successive time delay red-shift (z) measurements from the <u>same</u> light emitting galaxy. The data to be compared must have adequate time separation to provide enough red-shift change as calculated by expansion theorized 'Hubble Law', (H# = velocity / distance) to be measured within the red-shift calculated resolution of the spectrometer.

Red-shift measurements of what are now known as the ACO clusters were taken several decades ago (in the 1950's) by the Palomar All Sky Survey (PASS) shortly after the then world's largest telescope became operational. The Abell (ACO) cluster galaxies data were documented from photographic plates by Abell, Corwin, and Olowin. What needs to be done is measure the red shift for some of the same galaxies again every decade, and compare the results! This has been done once. Struble and Rood (S&R) cataloged data in 1999 from some of the same ACO galactic clusters and the data has been compared. The results show no velocity increase; i.e., <u>no acceleration</u>.

Galaxy clusters are used for comparing time lapse data. ACO # 2744 was used in the sample because time lapse data is available.

That cluster was also cataloged by S&R in 1999. ACO #2744 consists of 137 galaxies, is located in the southern hemisphere; RA 00.11.8; Dec -30.40 based on Sky Atlas version 1950.

The results indicate there is no measurable universe expansion. Acceleration is defined as the rate of change of velocity of a specific object (with respect to <u>time</u>, not distance). The previously named "New Universe Theory" is now shown to be the "Laws of Physics Model of the Universe".

Astronomers understand red-shift occurs when electromagnetic radiation emitted from an object is shifted towards the red end of the spectrum due to the objects receding velocity, known as the Doppler Effect. The amount of red-shift is a measure of the separation velocity between the observer in the Solar System in the Milky Way galaxy and some distant object.

Red-shift (Z) data also prove farther away galaxies are traveling faster than closer ones, thus by logic we know far away galaxies are separating from each other faster than closer ones. The astronomical observations made in the 1920's and 1930's correctly indicate farther away individual galaxies have higher separation speed with distance. But logic does <u>not</u> <u>support</u> <u>their</u> <u>assumed</u> conclusions that: 'Individual galaxy speeds are increasing with time! The universe is <u>not</u> expanding at an accelerating rate'. The BB theory was deducted from their presumption of expansion from a 'hot dense origin'. Their concepts violate several pertinent, proven, and never disproven laws of physics. (Listed in Figure 26)

Many in the astronomical community have generally accepted the notion that galaxies are accelerating apart, however, scientific logic must conclude that individual galaxy velocities are decreasing with time. All data used to show increasing velocity (even to date) was on different galaxies, not any specific object. Deceleration can and does occur by conversion of linear momentum to angular rotations.

How to prove there is no expansion acceleration.

At least several decades of time lapse are required between multiple galactic red shift observations of any individual galaxy to reveal either increasing or decreasing red shift. These 'Z' data will demonstrate one of <u>three possibilities</u>:

#1. <u>Increasing Z</u> with time; (Supports expansion theory);
#2. <u>No change in Z</u>. (Proving no velocity change and no universe expansion during this time increment);
#3. <u>Decreasing Z</u>. (Proving no expansion and galaxies' outward speeds are slowing with time).

Answers #2 and/or #3, or a mixture of #2 and #3, will prove the universe is <u>not</u> expanding! If the spectrometer resolutions are adequate, the results will demonstrate the universe is simply growing, (velocity is outward, but space is <u>not</u> expanding.) Change in distance is increasing, ($\Delta z/d$ is +), but change in red- shift and velocity are not changing with time ($\Delta z/t = 0$ or is -); because 'd' is the distance between different galaxies and; 't' is in reference to the same specific/ individual galaxy with time.

Abell (ACO) cluster catalog data is available from a few decades ago which can be used for base line t and Z data. ACO cluster red shift data ranges from .0316Z to .3080Z. Future red shift data compared to original Abell Cluster's alpha galaxies' Z's will prove velocity increase (acceleration) or decreasing velocity (deceleration) during several decades of time between data sets.

If after adequate time, we find no change in velocity or galaxies' velocities are decreasing (negative Δv), that will prove there is no expansion and was no BB. This has already been done, but the significance has not been exposed. Are the spectrometer data resolutions thought to be inadequate so that the results are thought to be inconclusive?

Looking for Proof ? ... There is no BB acceleration !

Repeat measurements of 583 ACO galactic clusters' red shifts were again catalogued in 1999 by Strubble and Rood [43] (first measured in the 1950s by Abell, Corwin & Olowin). Of these clusters 228 show <u>no</u> red shift change for four decades. The other cluster's alpha galaxies (most prominent galaxy reference) have random red shifts, apparently due to meandering and circulatory motions within those clusters because they are too far from the gravitational center of their host cluster. The 228 'no Z change' clusters strongly imply no increase in velocity (no expansion / acceleration). By using the following '5 step' analyses, we can verify <u>no expansion</u> and <u>H# is only a distance description</u>.

---

5 Step analyses: ** (Math Calculations are in Appendix 2)
<u>1</u>. Calculate the cluster velocity from the ACO red shift data.
<u>2</u>. Calculate the distance to the cluster. (Using the H# location description with velocity as calculated from Z value)
<u>3</u>. Calculate the distance traveled during the elapsed time and add the original distance to determine the new distance.
<u>4</u>. With the new distance and the H# distance formula, calculate the new velocity, as 'presumed' by the BB theory.
<u>5</u>. Compare the step 4 result with the more recent velocity as measured by the more recent (S&R) red shift observations.

** If you try some of the Appendix 2 calculations (and I hope you do), it is essential your calculator can handle at least 21 significant figures. Calculators TI-30Xa and HP equivalent lose digits in calculations. TI-84Plus is adequate.

Next question; "Is the time lapse between Z measurements adequate to mathematically confirm no Hubble expansion? Time lapse data with the '5 step' analyses indicates four decades of time is more than adequate if the red-shift data significant figure resolution is sufficient. (See Appendix 2) Several of the 228 need to be '5 step' analyzed to confirm data resolution requirement.

7C. Dark Matter (not dark energy)

A challenge to Nuclear physicists!

Dark mass has been shown to exist and is about eight times more plentiful than all of the other mass in the universe [39]. This is proven fact but the dark mass defies our ability to define it in detail. By logical deduction, it apparently is made of multiple neutron nuclides, which are called nucleonos.

Free neutrons in the universe have a half life of 10.25 minutes according to the "Chart of the Nuclides" [15]. I have the full chart on my office wall and frequently glance at the empty box (Z0, N2) at the left-bottom (beginning) of the chart. I often wonder if the empty box was placed there by a forward thinking scientist, or simply by the draftsman. Either way it is my belief that multiple neutron nuclides exist and are possibly the most stable form of matter in the universe.

So far, nucleonos have been observed only by indirect evidence, but have not been observed in the laboratory. We know invisible mass exists in and around galaxies, because without it, at their rotational angular velocities, the rotating galaxies would fly apart by centrifugal force. Thus, we have observed that 'dark matter' is invisible.

The invisible dark matter is likely to explain much of the observed lens effect as observed around distant galaxies. (usually referred to as 'gravitational lensing') In addition to gravitation effects on light waves, gravity holds and concentrates clouds of nucleonos around all objects. The result is a lens, not just a lens affect. The effect can be observed, but not the transparent lens itself.

A research challenge: Find a direct way (other than gravity) to observe nucleonos, singularly or in cloud-conglomerates. Since the clouds of nucleonos around galaxies vary in density, index of refraction does not appear to provide a solution. Refraction by gravity may be misconstrued to cause refraction? Nucleonos ...?

<u>Appendices Contents</u>

<u>Appendix 1.</u>

Several Extremely Large Telescopes* (ELT) are in the making to provide capability needed to research important questions. Planned ELT spectrograph resolutions for red shift measurements are not known?
*(ELT's diameters are 30+ Meters) [44]

<u>KECK TWINS</u>
10 Meter diameter, 2 each (2009 largest)
Operational since 1999
Mauna Kea, Hawaii

<u>LBT</u> (Large Binocular Telescope)
8.4 Meter, 2 each (equiv. 22.8 meters)
Operational since January 24, 2010
Mt Graham, SE Arizona, (10,480 feet)

<u>OWL</u> (Europe's Over-Whelming Large)
100 Meter diameter
Operational date: 2020?
Mauna Kea, Chili, or Canary Islands

<u>EURO50</u> – (TMT, Thirty Meter Telescope)
50 Meter diameter
Operational date: 2015
Chile or Canary Islands

<u>GMT</u> (Giant Magellan Telescope)
28.5 Meter
Operational date: 2016
Canary Islands?

<u>FAST</u> (Worlds largest single dish Radio Telescope)
500 Meter,
Operational date: 2014
Dawodang, Guizhou Province, (southwest) China

<u>LSST</u> Large Synoptic Survey Telescope
8.4 w/3200 mega pixel camera, 30 terabytes data/night
(World's largest)
Operational date: 2017
Cerro Pachon, Chile

Appendix 2

Proof of no expansion (Red-Shift explained, see Appendix.9)
Time lapse red-shift (ΔZ & Δv) analyses; Is the universe … ?
A. **Expanding (increasing in size at an accelerating rate)**
B. **Growing Consistent with Laws of Physics! Decelerations behind a constant speed (~c), outward progressing source …**

Figure # 21

Is Hubble Law just a description, not an equation? ☹

Five steps can validate or invalidate if Hubble Number (H = v / d) can or cannot be used as an equation.

(H # has been proved to be an approximate distance marker).

In 1999, Strubble and Rood [25] (S&R) re-cataloged Z values for 228 ACO (Abell, Corwin, Olowin) galactic clusters, originally cataloged in 1959. After several decades, data indicates no change in their recessional velocities. *Common measured clusters' Z values for both 1950's ACO and 1999 S&R cluster catalog (figures 22 & 23) Z's range from .03160 (ACO #1177) to .3080 (ACO #2744). (Data from catalogues, references* [36] *and* [37]*). Tentatively presuming H=v/d as an expansion equation, the following steps show velocity starting at v_1 increases over 40 years by Δv = .245708 Meters /second. S&R data with at least (?) four digits resolution indicates no velocity increase.* (Ref.: H# @ 75 Km/s/ Mps, ≈ H# @ 23,006 Km/s/ Bly; c = 299,792 Km/s; One Billion light years (Bly) = 9,468,376,000,000,000,000,000 Kilometers)

1. Calculate the cluster's v_1, using earliest catalog z_1 = .3080 for ACO #2744.
   $$v_1 = c \{[(z_1 + 1)^2 - 1] / [(z_1 + 1)^2 + 1]\} = 78613.80735 \text{ Km/sec}$$
2. Calculate (d_1) for z_1, using the Hubble 'equation';
   $$d_1 = v_1 / H\# = 3.418791324 \text{ Bly}$$
3. Calculate new distance (d_2 after time lapse years)
   $$d_2 = [d_1 + [(v_1) (\Delta t)]] = d_1 + 9.9248107940000 \times 10^{13} Km$$
4. Calculate Hubble presumed v_2' based on d_2 and H#
   $$v_2' = [(H\#) (d_2)] = 78613.80760 \text{ Km/sec}$$
5. Compare BB calculated v_2' with measured v_2 and v_1
   $$v_2' \neq v_1 : \Delta v = (v_2' - v_1) \text{ Per BB, } \Delta v = .000245708 \text{ Km/s}.$$

Measured velocity, v_2 & v_1, has not changed for decades; 1959 … 1999. But, ACO & S&R cataloged data show only 5 significant numbers. The question remains; Are 5 figures of accuracy and resolution adequate to confirm no velocity change? How much is required to prove | (v) ≠ (H#) (d) |? *Prove acceleration or no acceleration with Z data, spectrometer resolution needs 8 significant figures, e.g., z = .30800000 vs z = .3080, as listed in catalogues.*
 The necessary resolution is defined from the following exercise.

Five Steps

Appendix 2 (Continued)

Accelerating expansion is myth!
The 5 step analyses can be applied to any or all 228 S&R (Struble and Rood) repeat measurements of ACO Z values. Their actual time lapse repeat measured ΔZ's = 0.0000, implying their velocities did not change. The '5 step' will support or deny the increase in velocity as claimed by 'Hubble Law' if data resolution is adequate. ACO # 2744 was measured at a red shift of .3080 at both ends of a 40 year time lapse. "Hubble Law", if valid, by calculation claims the Galactic cluster's outward velocity increase (Δv, acceleration) is .245708 m/s after 40 years.
The equivalent 40 year average is about 6 cm per decade!

The H# calculated/claimed velocity increase, if it exists is below 1999 data resolution for Z measurements because; (1) the time lapse of 40 years is inadequate; (2), the red shift Z and wave length measurement resolutions are inadequate; or (3), there is no velocity increase and H# law is not valid as an equation (No BB)?

The exercise to determine the Z resolution necessary for measuring .245708 m/sec velocity at red shift of Z = .3080: Artificially increase the basic Z values by additions of one figure increments with recalculations for each v, until increment additions result in velocity change of $\geq$ .245708 m/s, the H Law claimed velocity increase = .000245708 Km/sec more than the velocity at Z = 3080.

The ΔZ accuracy for confirming or disproving the calculated Δv as presumed by H# expansion theory were determined to be seven figures to measure $\Delta v \leq$ .245708 m/s, (i.e., Z = .3080001). Data catalogued by S&R and ACO was listed at five for Z = .30800. Based on the reference value Z = .30800 for ACO # 2744, the H# calculated **Δv is +.245708 m/sec** after 40 years time lapse.

S&R and ACO data comparisons indicate no expansion. Proof of no expansion beyond doubt requires Z data measurements with resolutions of seven (7) significant figures to prove no Δv.

Appendix 2 (Continued)

Resolution & Accuracy!

Based on the reference value Z = .3080 for ACO #2744, the H# calculated Δv is +.00021 Km/sec after 40 years. To examine the resolution needed for confirming there is or is not accelerating expansion, hypothetical red-shift values beyond the four digit cataloged resolution of ACO and S&R red-shift data are assumed.

Figure # 22

The following is the exercise used to determine how much red shift (Z) measurement resolution is needed to prove the H# calculated Δv is, or is not valid.

Time lapse is 40 years (__1,262,476,800__ seconds, __c = 299,792__ K/s) H# calculated time lapse for __$\Delta v = .21$__* meters / sec. This defines the red shift resolution needed for confirming (v-v₁) = $\geq .21 m/s$.

Redshift (Z) when cataloged by ACO is .3080, and is equivalent to __v_1 = 78617.91__ Km / sec.

Z_2 $[(z+1)^2 - 1]$,   c x [ ]/[ ], $(v - v_1)$ Km/sec

__H# Calculated Δv after 40 yrs__ .... .00021

Presumed Z:	%c:	v after 40 years:	Δv:
.308000000	.71086400	78613.80736	**0.00000**
.308(1)		78635.14995	18.28475
.3080(1)	.71089016	78615.94174	2.13438
.3080(01)		78614.02080	.21344
.3080(001)		78613.82871	.02135
.3080(0001)		78613.80950	.002137959
.3080(00001)*	.710864003	78613.80758	**.000218640***
.3080(000001)**		78613.80736	**.000053331****

The above table lists presumed Z values after 40 years. The measured Z data must be accurate to the same number of significant figures with both the initial and post time lapse Z measurements (Numbers in parentheses in table).

__*Red shift measurement__ resolution of 9 digits at both ends of a 40 year time lapse, can prove there is no expansion, only growth consistent with the laws of physics (no acceleration). The double asterisk (**) calculation with new ELT telescopes (Appendix A) __and / or__ with only adequate spectrographic resolutions, proof can be confirmed after a new (only) 10 year time lapse study.

<u>Appendix 2</u> (Continued)

Request for study! (Plead!) ☺
<u>During the next 10 years, prove that the universe is only growing consistent with Laws of Physics (No acceleration)</u>

Current data presented in the "5 Step" indicate the Laws of Physics Model of the Universe is valid, but resolution of measurements are inadequate for conclusive proof of no galaxy acceleration. The task is defined for conclusively proving within one decade 'no expansion'. This uses H Law (H = v/d) to prove the H# is valid for estimating distance, but not velocity.

<u>First:</u> Determine from; "five step" the velocity with ten year time lapse, what Z (red shift) with five significant figures will calculate a measurable Z increase (if it occurs) using "H# Law".

<u>Next:</u> After the time lapse, re-measure the Z for the <u>same</u> galaxy to show no velocity and no red shift (Z) increase.

Z .... <u>.30800</u> (Galaxy cluster ACO #2744)
v .... **78613.80735 Km/sec**
d .... **3.1 x 10 ^{22} Km**
<u>40 yr Δd</u> ... **9.924810794 x 10 ^{13} Km**
        $d+ \Delta d$ ...... **7.443450878924810795000 x 10 ^{22} Km**
v_2 .... **78613.80760 Km/sec**
Δv .... **.000245708 Km/sec**

Z .... <u>.50001</u> (Galaxy target to be determined)*
v .... **115306.3183 Km/sec**
d .... **4.747908758 x 10 ^{22} Km**
<u>10 yr Δd</u> ... **3.639288794 x 10 ^{13} Km**
$d+ \Delta d$ ...... **4.7479087616392887940000 x 10 ^{22}Km**
v_2 .... **115306.3190 Km/sec**
Δv .... **.0007 Km/sec**

The two cases cited compare higher Z value and lower time lapse. The higher Δv indicates the required red shift resolution to prove no acceleration will be reduced by a factor of three even though the time lapse has reduced by a factor of four.

*Selected galaxy for time lapse must be near its cluster's C of G.

Appendix 3

Glossary

Scientific thinking uses logic with <u>only</u> proven facts and verified phenomena. Three inherent traits/obstacles we humans must overcome, to advance knowledge of the universe beyond myth and fiction:

Prejudice: Any preconceived opinion or feeling, either favorable or unfavorable.

Bigotry: Stubborn intolerance of any belief, or opinion, that differs from one's own.

Dogmatism: An authoritative, arrogant assertion of unproved or unprovable principles.

Big Bang theory (**BB**): A concept based on the (<u>mis</u>?) interpretation of red-shift observations indicate that galaxies are accelerating apart in all directions. (Without force!) For the origin of the BB universe … it is asserted it all erupted out of a single point smaller than a speck of dirt under your fingernail!

Model of the Universe with Laws of Physics (**LOP Model**); Now, with proof, this concept can be recognized as <u>Universe Model with Laws of Physics</u>. This concept for origin of the universe is based on scientific thinking; it is plausible because it uses logic with <u>only</u> proven facts and verified and repeatable phenomena (universal laws of physics).

Abell (ACO) Cluster: A group of 30 or more galaxies that are orbiting about each other, together in their mutual gravity fields. Some have several hundred galaxies. Abell cataloged the first 2027. Abell died in 1959, and Harold Corwin and George Olowin finished the catalog of 4076. They are now appropriately known as the ***ACO Clusters***.

Accretion: An increase in the mass of a celestial object by the collection of surrounding interstellar gases and objects by gravity.

Annihilation: Process in which a particle and an antiparticle unite and obliterate each other, with their masses transformed into electromagnetic energy in the form of photons.

Atomic number: The number of positive charges or protons in the nucleus of an atom.

Black Hole: An area of space-time with a gravitational field so intense that its escape velocity is equal to or exceeds the speed of light. Succinctly, nothing can escape from a black hole's gravitational sink.

Chandrasekhar Limits: Size range for a star that can evolve into a white dwarf star. The limits are .9 to 1.44 times the mass of our sun. These limits are named for the Indian-born American astrophysicist first to do the math.

Cluster: A group of the same or similar objects gathered or occurring closely together; a bunch

Corporeal: Real and tangible. Includes mass and energy

Cosmology: The study of the physical universe considered as a totality of phenomena in time and space.

Dark Energy: The term used for the hypothetical energy/phenomena to power the idea of an expanding universe and to rationalize the (mis)interpretation of red-shift data as indication of accelerating dispersal of galaxies. The farther away a galaxy the higher the red-shift was first observed in 1910 by V M Slipher, but red shift can only provide a measure of velocity, not acceleration.

Dark Matter: The mass that is needed in and around galaxies to provide the gravity needed to balance the centrifugal force from the observed rotational angular velocity of the galaxy.
The dark matter mass needs to be 6 to 10 times more than all visible mass combined. See *Nutralinos*.

Deflagration: Propagation of flame through a combustible mixture. A chemical term borrowed for the New Universe Theory which involves spherical propagation of chain reactions of positronium annihilations as the deflagration front progresses into the primordial matter outside the universe.

Descendant: Something derived from a prototype or earlier form.

Electromagnetic Spectrum: The distribution of energy emitted by a radiant source, as by an incandescent body, arranged in order of wavelengths.

Electron: A stable subatomic particle in the lepton family having a rest mass of 9.1066×10^{-28} grams and a unit negative electric charge of approximately 1.602×10^{-19} coulombs.

Electron Volt: A unit of energy equal to the energy acquired by an electron falling through a potential difference of one volt. One million electron volts (Mev) = ~ 44.5 million trillionths of a kilowatt hour.

Element: A substance composed of atoms having an identical number of protons in each nucleus. The atoms of elements have one electron for each proton in the nucleus, orbiting the nucleus at various specific distances. The nucleus is the nuclide of the atom/element. There are 112 known elements (106 in the periodic table + 6 artificially produced).

Energy: A form of matter since both mass and energy can be converted to and from the other by Einstein's relationship $E=mc^2$; Also, energy is the capacity of a physical system to do work;

Energy exists in many forms (e.g., electromagnetic radiation, heat, mechanical work, etc).

Globular Cluster: A system of stars, generally smaller in size than a galaxy, that is more or less globular in conformation.

Gluon: Hypothetical massless, neutral elementary particles believed to mediate the strong interaction that binds quarks together.

Hexahedron Crystal: A homogenous six sided solid formed by a repeating, three-dimensional pattern of atoms, ions, or molecules and having fixed distances between constituent parts.

Hubble Number (H#): A ratio expressing the rate of apparent expansion of the universe, equal to the velocity at which a typical galaxy is receding from Earth divided by its distance from Earth.

Mass: A form of matter since both energy and mass can be converted to and from the other by the relationship $E=mc^2$. Mass has a property equal to the measure of an object's resistance to changes in either the speed or direction of its motion. The mass of an object is not dependent on gravity and therefore is different from but proportional to its weight. In nuclear physics mass of elementary particles are often expressed in equivalent energy units.

Matter: Something that has mass and exists as a solid, liquid, gas, or plasma; also in the form of energy. Since mass can exist in the form of energy by Einstein's formula $E=mc^2$, matter includes both mass and energy. By the 1st law of thermodynamics, matter cannot be created or destroyed.

Model: A schematic description of a system, theory, or phenomenon that accounts for its known or inferred properties and may be used for further study of its characteristics.

Nebula: A diffuse mass of interstellar dust, or gas, or Supernova debris, or all three, visible as luminous patches or areas of darkness depending on the way the mass absorbs or reflects incident radiation.

Neutron: The neutrally charged particles in the nucleus of atoms, each containing three quarks, and gluons. Rest life of a single free neutron is 10.25 minutes, at which time it decays into a proton and an electron (beta particle) plus a gluon.

Neutron Star: A celestial body consisting of the super-dense remains of a massive star that has collapsed with sufficient force to push all of its electrons into the nuclei that they orbit, thus leaving only neutrons, and having a powerful gravitational attraction from which only neutrinos and high-energy photons can escape, rendering the body detectable only by x-ray.

Nuclide: The nucleus of atoms. Chart of the nuclides is a graphical listing of all known atomic nuclei, listed by the number of protons, compared to the number of protons + neutrons. More than 3600 have been discovered.

Nutralinos: Yet to be specifically observed, but are theorized to be the massive stable neutral particles, each made up of two or more neutrons but no protons. They are theorized to be the dark matter sometimes referred as WIMPs, (weakly interacting massive particles) that are needed to provide the gravity to balance the centrifugal force and prevent the galaxies' stars from flying apart. See *Dark Matter*.

Open Cluster: A group of stars orbiting in the galactic disk. (e.g., Pleiades, Bee Hive,)

Parsec: A unit of astronomical length based on the distance from Earth at which stellar parallax is one second of arc and equal to 3.258 light-years, 3.086×10^{13} kilometers, or 1.918×10^{13} miles.

Photon: Quantity of electromagnetic radiation ... usually considered an elementary particle that is its own antiparticle and has zero rest mass and no electrical charge.

Planetary Nebula: The ring shaped nebula around a newly formed white dwarf star. The nebula is the material that has been 'shrugged-off' by a sol sized star that has recently completed its fusion life and completed its 'swollen' red giant dying phase.

Positron: A stable subatomic particle in the lepton family having a rest mass of 9.1066×10^{-28} grams and a unit negative electric charge of approximately 1.602×10^{-19} coulombs. An elementary particle having the same mass as an electron, but having a positive charge equal in magnitude to an electron; the anti-particle of an electron.

Positronium: A sub atomic system composed of a mutually orbiting electron and positron pair; held together by their electrical charge, but separated by their mutually orbiting centrifugal forces. Usually short lived unless stabilized by other forces. When annihilated, two 511,000 ev gamma ray photons are radiated.

Primordial: First existing ... primordial matter.

Progenitor: A direct ancestor. An originator of a line of descent; a precursor

Proton: The positive charged particle in the nucleus of atoms, and contains three quarks and gluons.

Quark: The sub-elementary particles of protons and neutrons. There are six types, often referred to as flavors.

Quark Star: A hypothetical celestial object that is the remnant of a massive star that has collapsed with a force sufficient to reduce all particles to strange quarks. Also called strange star. Two probable candidates have been observed as of 2006.

Red-Shift: An increase in the wavelength of radiation emitted by a celestial body as a consequence of the Doppler Effect. Doppler Effect is the lengthening of radiating (light) waves as a result of separation velocity between the source and the observer location. Red shift is

quantified in the range of zero to infinity, notated as 'Z'. Z is equal to the apparent wave length less the at-rest wave length of similar element line spectra, divided by the at-rest wave length.

***Shwarzschild Radius*:** The radius within which if an object collapses and light can no longer escape the gravitational field, thus, the object becomes a black hole.

***Supernova*:** A rare celestial phenomenon involving the explosion of most of the material in a star, resulting in an extremely bright, short-lived object that emits vast amounts of energy. Supernovae are so bright that, for a short time of about a few weeks, are as bright as their host galaxy.

***Supernovae Type 1a (also referred to as Se1a, Se 1a, Se Ia, Type Ia)*:** The explosion of a dormant white dwarf star, which formerly was a star with .9 to 1.44 times the mass of our sun, and after using so much of its mass as nuclear fuel can no longer fuse nuclides and produce energy... The white dwarf, subsequent to progenitor burn-out, accreted material from a neighbor star and again reached critical mass. The brightness of all Type 1a are consistent in intrinsic brightness and therefore provide an accurate way to calculate their distance from us. It is thought further study of Se1a supernovae will allow more refinement of distance calculations by further 'calibrating brightness and spectral content'.

***Transmuting*:** This is the changing of a substance from one form into another.

Universe: All matter, Mass and energy, including the earth, the galaxies, and the contents of intergalactic space, regarded as a whole.

Vortex: A mass of objects spiraling inward or outward, whirling with a circular motion.

White dwarf: The remnant of a .9 to 1.44 sol mass star that has consumed so much of its nuclear fuel that its total mass is insufficient to provide the self gravity for maintaining the internal pressure and density needed to sustain nuclear fusion processes. It has collapsed into an extremely dense state with no empty space between its atoms, but not reaching the extremely denser state of a neutron star, quark star, or black hole.

Appendix 4

References and Credits

[1] "New Universe Theory with Laws of Physics", ISBN: 1-4184-9430-5 (dj) © 2005 Bobby McGehee Library of Congress control Number: 20055904178

[2] <http://newuniversetheory.com> page 4, First publication of the "Universe Milestone Bar- Chart" (2005)

[3] John Huchra home page: Hubble number listings 1928 thru 2007. <http://www.cfa.harvard.edu/~huchra/>

[4] First estimate for our distance to the center of the universe; Figure 7.7 page 162 of Ref [1]

[5] Michael Richmond's Supernova Page; Sne after 1988 to 4.16.2002, <www.tass-survey.org/richmond/sne/sne.org>

[6] "Type Ia Supernova Discoveries at z = 1 From the Hubble Space Telescope: Evidence for Past..." 2004 arXiv:astro-ph/0402512v2: Adam G Reiss, Louis-Gregory Strolger, John Tonry, Stefano Casertano, Henry C Ferguson, Bahram Mobasher, Peter Challis, Alexei V Filippenko, Saurabh Jha, Weidong Li, Ryan Chomock, Robert P Kirshner, Bruno Leibundgut, Mark Dickerson, Mario Livio, Mauro Giavalisco, Charles C Steidel, Narciso Benitez, Zlatan Tsvetanov

[7] "New Hubble Space Telescope Discoveries of Type Ia Supernovae at Z >1 ..." 2007 arXiv:astro-ph/0611572v2: Adam G Reiss (JHU, STScI), Louis-Gregory Strolger (UWK), Stefano Casertano (STScI), Henry C Ferguson (STScI), Bahram Mobasher (STScI), Ben Gold (JHU), Peter J Challis (CfA), Alexei V Flippenko (UCB), Saurabh Jha (UCB), Weidong Li (UCB), John Tonry (IfA), Ryan Foley (UCB), Robert P Kirshner (CfA), Mark Dickinson (NOAO), Emily MacDonald (NARO), Daniel Eisenstein (UofA), Mario Livio

(STScI), Josh Younger (CfA), Chun Xu (STScI), Tomas Dahlen (STScI), Daniel Stern (JPL)

[8] Bootes Void
< http://www.acceleratingfuture.com/michael/blog/?p=69 >

[9] "Extragalactic … and the WMAP Cold Spot"; 'Largest Void' (arXiv 0704.0908v2 [astro-ph] 3 Aug 2007); by Lawrence Rudnick, Shea Brown, Liliya R Williams, Department of Astronomy, University of Minnesota.

[10] British UK team Jack Milton (UK), Kate (UK), Issak (Clavendish Laboratory U of Cambridge), Robert Priddey (Imperial College in London), and Richard McMahon U of Cambridge) reports in April 2003, the Most distant Quasar Highest Red-Shift (Z = 6.43) Farthest Quasar ever at 87% of c.

[11] Anna Frebel, McDonald Observatory, University of Texas, (Quasar pg 7 ?) Heads team studying ages of ancient stars; press release May 2007, discusses 13.2 billion year old stars, star # HE 1523-0901. SpaceRef.com & Universetoday.com

[12] "Introduction to Fluid Mechanics" by Stephen Whitaker, 1968 Krieger Publishing Co., Malabar Florida

[13] "Mechanical Engineers Handbook", Lionel S Marks, McGraw-Hill Book Company.

[14] Catalog of Abell Clusters by George Abell (Cal Tech), Harold Corwin (Cal Tech, formerly University of Edinburg), and Ronald P Olowin (University of Oklahoma)

[15] "Chart if the Nuclides" Published by Knolls Atomic Laboratory of the Lockheed-Martin Corporation. 16th ed.,
Web: <ChartOfTheNuclides.com> Edward M Baum, Harold D Knox, Thomas R Miller, (KAPL) (Also, See 'Chart of Nuclides' on <nndc/bnl.gov>).

[16] First technical analysis/description of the 'Big Bang Model'; "First three Minutes", by Steven Weinberg; first printing February 1977; published by Bantam/Basic Books, ISBN 0-553-14131-7.

[17] BB model "History of the Universe Poster". By Particle Data Group of Lawrence Berkeley National Laboratory <http://particleadventure.org/>

[18] NUT New Concept Model Deflagration Wave from Reference [1] Figure 6.14

[19] Beijing Electron Positron Collider (BEPC); Web site: <www.ihep.ac.cn/English/E-BEPC/index.htm>

[20] COBE < http://www.britannica.com/eb/topic-139203/Cosmic-Background-Explorer > Britannica Book of the Year 2007

[21] "Black Body Radiation", Michael Fowler, University of Virginia , September 7, 2008

[22] Planck's Law of Black Body Radiation, < http://en.wiki/Planck's_law_of_black-body_radiation >

[23] "WMAP Glossary of Technical Terms" < http://map.gsfc.nasa.gov/m_help/h_glossary.html >

[24] M4 White Dwarf Stars Cool, Harvey Richer (UBC), NASA <http://antwrp.gsfc.nasa.gov/apod/ap971102.html>, also, <http://antwrp.gsfc.nasa.gov/apod/ap950910.html>

[25] SETI is the name used for several projects searching out 'Extra-terrestrial Intelligence'. The first private SETI organization was an off-shoot from the 'Planetary Society' which was founded by Carl Sagan, Bruce Murray, and Louis Friedman, in 1980. (I was a charter member). SETI@home efforts are now coordinated by SSL (Space Science Laboratory) which was initiated in 1958 at UC at Berkeley.

[26] International <u>Supernovae</u> Network:
< http://www.supernovae.net/isn.htm >;
Michael Richmond; web site maintained by David Bishop

[27] Science, publication of American Association for Advancement of Science (AAAS) published a solicitation for internet users to help classify galaxies catalogued from the Sloan Digital Sky Survey (SDSS) telescope in Sunspot, NM. The web site is Galaxy Zoo < www.galaxyzoo.org >.

[28] List of Abell clusters (includes all ACO clusters)
< http://en.wikipedia.org/wiki/Abell_catalogue >

[29] Web sites; www. Chemical elements . com (delete spaces to link)

[30] "History of the Origin of the Chemical Elements ...", 2004, Norman E. Holden, Brookhaven National Laboratory, Upton NY 11973-5000, USA

[31] "History of the Universe Poster", presented by the Particle Data Group of Lawrence Berkeley National Laboratory (LBNL) <particleadventure.org/frameless/history-universe.html>

[32] "Galileo's Daughter" by Dava Sorbel; 1999, ISBN 0 14 02.8055.3

[34] Wil Tirion "Sky Atlas 2000.0"; Sky Publishing Corporation, Cambridge Massachusetts & Cambridge University Press, London, New York, New Rochelle, Melbourne, Sydney.

[35] "Cosmos" by Carl Sagan, Random House Inc., 1983, ISBN 0-394-50294-9 and ISBN 0-394-71596-9

[36] Strasbourg Astronomical Data Center (CDS). The Smithsonian/ NASA Astrophysics Data System, 1972

[37] Steven Coe, Astronomy Magazine March 2009, Article; "Discover galaxy groups and clusters".

[38] Nasa/ipac Extragalactic Database (NED) publication "1999 ApJS...125...35S", posted by Struble and Rood.

[39] "The Hunt for Dark matter in Galaxies", Ken C Freeman, Mount Stromlo Observatory, Australia National University, Weston Creek, ACT 2611 Australia, Published AAAS journal "Science" 12 December 2003.

[40] "Before the Big Bang", Ernest J Sternglass, by Four Walls Eight Windows, 39 W 14th Street, New York, NY published 1997.

[41] Paul Adrien Maurice Dirac was born on 8th August, 1902. Dirac's work has been concerned with the mathematical and theoretical aspects of quantum mechanics. He wrote a series of papers on the subject, published mainly in the Proceedings of the Royal Society, leading up to his relativistic *theory of the electron* (1928) and the theory of holes (1930). This latter theory required the existence of a positive particle having the same mass and charge as the known (negative) electron. Thus, the positron was discovered experimentally at a later date (1932) by C. D. Anderson, while its existence was likewise proved by Blackett and Occhialini (1933) in the phenomena of "pair production" and "annihilation".

[42] NED ...NASA / IPAC Extragalactic Database...including the CDS which currently has on file about 8232 catalogues of stars, galaxies, and extragalactic objects including the Abell (ACO). The Strasbourg Astronomical Data Center (CDS) [36] was created in 1972 to collect data concerning astronomical objects, making it available on electronic form.

[43] Figure 22 and Figure 23; Findings by author: ... 228 ACO (Abell, Corwin, and Olowin) galactic clusters searched from NED and CDS that were again cataloged decades later in 1999 by Strubble and Rood. These 228 clusters were found to have zero change in red shift (Z) during the intervening time increment.

[44] Information about Very Large Telescopes (VLT); presented in July 7, 2007 Astronomy magazine. Christina R Dunn is an astronomer at University College, London, and guest researcher at Lawrence University in Appleton, Wisconsin.

<u>Appendix 5</u> Data

Reference Data: Abell, Corwin, Olowin; ACO clusters

From the Abell Galaxy Cluster (ACO) Catalog

Figure # 23

Drag & Drop from http://webviz.u-strasbg.fr/vis-bin/VizieR4
VizieR Search Page VII/110A/table3

Sort for highest 15 ct

Abell, George; California I of Technology					
Olowin, Ronald P.; University of Oklahoma					
Corwin, Harold G.; University of Edinburg					
<u>Full</u>	<u>RA</u>	<u>Dec</u>	<u>Count</u>	<u>z</u>	<u>m10</u>
ACO #	"h:m:s"	"d:m:s"	<u>ct</u>		<u>mag</u>
1571	12 31.5	+83 37	190	0.209	17.6
586	07 29.1	+31 44	190	0.171	17.4
2721	00 03.6	-35 00	192	0.114	16.8
1413	11 52.8	+23 39	196	0.1427	17.1
1758	13 30.5	+50 46	198	0.28	18
2645	23 38.8	-09 19	205	0.246	18
777	09 22.4	+78 27	210	0.224	17.5
2218	16 35.7	+66 19	214	0.171	17.7
910	09 59.1	+67 24	222	0.2055	17.5
1146	10 58.9	-22 27	222	0.141	17
3558	13 25.1	-31 14	226	0.0482	14.7
1689	13 09.0	-01 06	228	0.181	17.6
2125	15 40.5	+66 28	230	0.2465	17.6
545	05 30.0	-11 34	234	0.154	17
665	08 26.2	+66 03	321	0.1816	17.5

1950 reference catalog was used, but for convenience, the plots of ACO clusters are plotted in the later 2000.0 Atlas reference frame. The difference is relatively small and is of course largely due to our (solar system's) changing position as we revolve around in the Milky Way galaxy.

Data

References: Abell, Corwin, Olowin
Clusters with no change for about four decades
Z data by Abell, Corwin, Olowin and again by
Strubble and Rood in 1999 (ACO# 1 thru #1548)

Figure # 24

Figure 24 cmn z 1-410a.xls					
ACO and S&R Galactic Clusters with common Z's					
Cataloged in 1959 and 1999 = ~ 40 yeqr time lapse					
			max=321	max=.308	max=19.6
ACO #	RAB1950	DEB1950	Count	z	m10
	"h:m:s"	"d:m:s"	ct		mag
1	00 05.0	+16 14	51	0.1249	17.1
23	00 19.2	-01 10	45	0.1052	17
24	00 19.9	+23 01	127	0.1338	17.5
43	00 26.3	+17 18	37	0.1114	15.9
71	00 35.1	+29 19	30	0.0724	15.5
75	00 37.2	+20 59	42	0.0626	15.5
84	00 39.2	+21 08	76	0.103	17.6
96	00 43.7	+39 14	61	0.1344	17.4
115	00 53.3	+26 03	174	0.1971	17.3
116	00 53.3	+00 22	48	0.0665	15.7
117	00 53.5	-10 18	40	0.0535	16
134	01 00.5	-02 48	43	0.0699	16
136	01 01.4	+24 48	99	0.1569	17.5
141	01 03.2	-24 52	140	0.23	17.7
158	01 09.1	+16 37	46	0.0645	15.9
160	01 10.2	+15 15	34	0.0447	15.7
168	01 12.6	-00 01	89	0.0452	15.4
171	01 14.1	+16 00	42	0.0706	15.9
180	01 19.3	+02 45	33	0.135	17
186	01 20.2	-10 41	56	0.1029	17.2
188	01 20.3	-13 02	57	0.123	17.2
209	01 29.5	-13 50	158	0.206	17.8
223	01 35.5	-13 02	152	0.207	17.6
224	01 35.8	-07 12	75	0.1617	17
234	01 38.3	+18 40	76	0.1731	17.9
236	01 38.0	-12 06	58	0.1874	17.2
243	01 40.0	-10 29	47	0.1117	16.6
245	01 41.5	+06 08	40	0.079	16.4
246	01 42.1	+05 33	56	0.07	16.4
272	01 52.4	+33 42	52	0.0877	16.8
279	01 53.8	+00 49	70	0.0797	17.2
281	01 54.6	-06 05	47	0.088	17
403	02 56.6	+03 18	100	0.1033	17.5
410	03 01.3	+03 36	70	0.0897	16.9

Data

Reference: Abell, Corwin, Olowin

Clusters ΔZ = 0. No change for about four decades.

Figure # 25

Figure 25 cmn z 450-1003.xls					
ACO and S&R Galactic Clusters with common Z's					
Cataloged in 1959 and 1999 = ~ 40 yeqr time lapse					
			max=321	max=.308	max=19.6
ACO #	RAB1950	DEB1950	Count	z	m10
	"h:m:s"	"d:m:s"	ct		mag
450	03 38.7	+23 20	40	0.0607	16.4
468	03 49.8	+21 16	34	0.1325	17.4
484	04 13.9	-07 47	50	0.0386	16.9
506	04 40.9	-09 48	88	0.1561	17.5
508	04 43.3	+01 55	85	0.1479	17.4
509	04 45.1	+02 12	72	0.0836	17.4
513	04 45.8	-09 48	32	0.1491	17.5
527	05 04.0	+73 38	34	0.0794	15.7
528	04 56.9	-09 05	40	0.2896	17.5
545	05 30.0	-11 34	234	0.154	17
562	06 46.5	+69 20	70	0.11	17
586	07 29.1	+31 44	190	0.171	17.4
588	07 33.6	+70 04	78	0.16	17.1
593	07 46.2	+72 57	154	0.226	17.4
595	07 45.0	+52 12	45	0.0666	15.6
639	08 15.1	+68 04	135	0.291	17.7
644	08 15.0	-07 26	42	0.0704	16.2
680	08 32.1	+37 02	61	0.079	17.7
690	08 36.2	+29 01	52	0.0788	16.9
732	08 55.3	+03 22	65	0.203	17.7
733	08 57.6	+55 49	64	0.1159	17.7
762	09 13.9	+74 30	32	0.1332	16.2
786	09 23.7	+75 01	45	0.1241	16.9
787	09 23.5	+74 37	106	0.1352	16.9
788	09 23.3	+72 31	63	0.1352	17
801	09 25.2	+20 47	81	0.1918	17.7
819	09 29.6	+09 53	36	0.0759	16.5
868	09 43.0	-08 25	186	0.153	17.6
873	09 48.4	+71 32	133	0.182	17.4
895	09 53.5	+49 44	53	0.36	18
910	09 59.1	+67 24	222	0.2055	17.5
923	10 03.7	+26 09	50	0.1162	17.2
924	10 04.1	+35 54	75	0.0989	17.2
1003	10 22.0	+48 03	37	0.052	16.6

Data

Reference: Abell, Corwin, Olowin
Clusters ΔZ = 0. No change for about four decades.

Figure # 26

Figure 26 cmn z 1073-1399.xls					
ACO and S&R Galactic Clusters with common Z's					
Cataloged in 1959 and 1999 = ~ 40 yeqr time lapse					
			max=321	max=.308	max=19.6
ACO #	RAB1950	DEB1950	Count	z	m10
	"h:m:s"	"d:m:s"	ct		mag
1073	10 39.6	+36 54	82	0.139	17.2
1093	10 44.3	+09 20	51	0.226	17.8
1094	10 44.8	+27 47	83	0.2004	18
1132	10 55.3	+57 03	74	0.1363	17
1149	11 00.4	+07 54	34	0.071	16
1170	11 04.9	+08 17	104	0.162	17.6
1177	11 06.8	+21 58	32	0.0316	15.7
1178	11 07.1	+34 52	103	0.2596	17.8
1201	11 10.4	+13 42	103	0.1688	17
1216	11 15.2	-04 12	57	0.0524	16
1224	11 18.2	+36 42	62	0.2897	17.8
1227	11 18.8	+48 18	112	0.112	16.6
1234	11 19.8	+21 40	88	0.1663	17.3
1235	11 20.3	+19 54	122	0.1042	17
1276	11 27.4	+33 18	54	0.0603	17.6
1278	11 27.4	+20 45	151	0.129	17.3
1279	11 28.3	+67 30	32	0.129	16.5
1292	11 29.1	+36 06	104	0.2319	17.5
1299	11 29.7	+34 15	104	0.2247	17.5
1304	11 30.1	+35 44	80	0.2131	17.5
1337	11 36.8	+10 26	50	0.0826	17.2
1342	11 38.1	+10 21	53	0.1061	17.2
1343	11 38.6	+60 56	32	0.1318	17.2
1345	11 38.6	+10 58	71	0.1095	17.2
1354	11 39.6	+10 26	57	0.1178	17.2
1356	11 39.9	+10 43	77	0.0698	17.2
1360	11 40.5	+11 18	66	0.1535	17.2
1372	11 42.9	+11 48	70	0.1126	17.2
1373	11 42.9	-02 07	94	0.1314	17.2
1374	11 43.3	+50 01	54	0.1314	17.4
1377	11 44.3	+56 01	59	0.0514	15
1382	11 45.6	+71 43	57	0.1053	15.9
1385	11 45.5	+11 50	52	0.0831	17.2
1399	11 48.6	-02 49	82	0.0913	16

Data

Reference: Abell, Corwin, Olowin
Clusters ΔZ = 0. No change for about four decades.

Figure # 27

Figure 27 cmn z 1400-1873.xls					
ACO and S&R Galactic Clusters with common Z's					
Cataloged in 1959 and 1999 = ~ 40 yeqr time lapse					
			max=321	max=.308	max=19.6
ACO #	RAB1950	DEB1950	Count	z	m10
	"h:m:s"	"d:m:s"	ct		mag
1400	11 48.8	+55 23	38	0.0778	17.2
1401	11 49.5	+37 33	153	0.1648	17
1402	11 49.9	+60 42	34	0.1648	17.2
1412	11 53.1	+73 45	86	0.0839	15.9
1413	11 52.8	+23 39	196	0.1427	17.1
1430	11 56.9	+50 04	96	0.2105	17.6
1452	12 01.1	+52 01	46	0.0631	15.7
1468	12 03.1	+51 42	50	0.0844	16
1495	12 10.9	+29 31	123	0.1429	17
1496	12 10.9	+59 33	58	0.0941	16
1497	12 11.6	+26 56	101	0.1669	17.8
1504	12 12.8	+27 48	98	0.1836	17.6
1514	12 15.4	+20 56	132	0.1995	17.6
1524	12 19.2	+08 07	103	0.1369	17.2
1525	12 19.5	-00 52	186	0.259	18
1548	12 26.5	+19 42	155	0.1611	17.5
1550	12 26.8	+47 59	167	0.254	17.8
1553	12 28.3	+10 51	100	0.1652	17.8
1609	12 44.0	+26 42	56	0.0891	16.8
1616	12 45.4	+55 19	39	0.0833	16
1622	12 47.3	+50 06	96	0.2855	18
1632	12 50.6	+29 05	80	0.1962	17.2
1643	12 53.6	+44 21	50	0.1981	17.7
1675	13 02.9	+34 49	50	0.184	17.2
1677	13 03.5	+31 10	112	0.1832	17.7
1679	13 04.2	+32 04	115	0.1699	17.5
1685	13 06.5	+35 01	44	0.197	17.2
1759	13 31.6	+20 30	132	0.168	17.6
1760	13 31.7	+20 28	168	0.1711	17.2
1774	13 39.0	+40 16	81	0.1691	17.6
1785	13 42.5	+38 24	90	0.2136	17.2
1836	13 59.0	-11 22	41	0.0363	15.7
1840	13 59.3	+30 49	35	0.1104	17.2
1873	14 09.5	+28 23	41	0.0776	16.3

Data

Reference: Abell, Corwin, Olowin
Clusters ΔZ = 0. No change for about four decades.

Figure # 28

ACO #	RAB1950	DEB1950	Count	z	m10
Figure 28 cmn z 1878-2179.xls					
ACO and S&R Galactic Clusters with common Z's					
Cataloged in 1959 and 1999 = ~ 40 yeqr time lapse					
			max=321	max=.308	max=19.6
	"h:m:s"	"d:m:s"	ct		mag
1878	14 10.6	+29 27	56	0.254	17.5
1880	14 10.9	+22 38	67	0.1413	17.2
1889	14 14.5	+30 57	112	0.186	17.3
1899	14 19.0	+17 55	33	0.0536	16
1904	14 20.3	+48 47	83	0.0708	15.6
1911	14 22.4	+39 11	80	0.1913	17.2
1914	14 24.0	+38 03	105	0.1712	17.2
1920	14 25.7	+56 00	103	0.131	17
1921	14 27.0	+23 19	63	0.1352	17.2
1929	14 29.7	+29 45	95	0.2191	17.3
1930	14 30.5	+31 50	60	0.1313	17
1936	14 32.9	+55 02	69	0.1386	17
1940	14 33.9	+55 22	130	0.1396	17
1942	14 36.1	+03 53	138	0.224	17.5
1952	14 38.9	+28 51	107	0.248	18
1954	14 39.9	+28 44	120	0.181	17.6
1957	14 41.0	+31 25	166	0.241	17.8
1961	14 42.4	+31 24	137	0.232	17.8
1979	14 48.9	+31 29	108	0.1687	17.2
1990	14 51.6	+28 17	140	0.1269	17.2
2020	15 01.3	+08 07	47	0.0578	16
2036	15 09.2	+18 15	39	0.1163	16
2050	15 13.8	+00 17	50	0.1183	17.1
2053	15 14.7	+00 30.8	75	0.1127	17.4
2064	15 19.4	+01 13	101	0.0721	17.2
2069	15 21.9	+30 04	97	0.116	16.6
2084	15 28.2	+35 28	57	0.342	17.3
2100	15 34.5	+37 48	138	0.1533	17
2108	15 37.8	+18 03	45	0.0919	15.7
2111	15 37.7	+34 34	148	0.229	17.8
2125	15 40.5	+66 28	230	0.2465	17.6
2158	16 06.6	+43 08	45	0.1349	16.8
2178	16 19.4	+24 46	51	0.0928	17.1
2179	16 18.6	+42 32	52	0.136	17.1

Data

Reference: Abell, Corwin, Olowin
Clusters ΔZ = 0. No change for about four decades.

Figure # 29

Figure 29 cmn z 2183-2616.xls					
ACO and S&R Galactic Clusters with common Z's					
Cataloged in 1959 and 1999 = ~ 40 yeqr time lapse					
			max=321	max=.308	max=19.6
ACO #	RAB1950	DEB1950	Count	z	m10
	"h:m:s"	"d:m:s"	ct		mag
2183	16 19.9	+42 50	56	0.1365	17.1
2198	16 26.5	+43 56	85	0.0798	17.7
2210	16 32.3	+05 35	50	0.1465	17.1
2220	16 38.5	+53 51	42	0.1106	17.5
2235	16 53.3	+40 06	73	0.1511	17.1
2240	16 54.0	+66 49	165	0.138	17.4
2246	17 00.4	+64 17	146	0.225	17.6
2250	17 09.1	+39 45	52	0.0654	16.5
2252	17 12.6	+49 27	63	0.1147	17.5
2257	17 16.1	+32 38	61	0.1054	17.1
2263	17 21.2	+26 59	46	0.1051	16.9
2265	17 13.7	+77 29	47	0.1051	17.4
2270	17 26.3	+55 13	49	0.2377	17.7
2283	17 44.9	+69 40	65	0.183	17.4
2301	18 15.2	+69 38	34	0.0874	15.8
2317	19 08.5	+68 59	186	0.211	17.6
2320	19 17.0	+70 54	85	0.171	16.9
2328	20 45.4	-18 00	81	0.147	16.4
2330	20 54.9	-22 14	91	0.1138	17.4
2339	21 18.3	-21 40	76	0.1128	17
2344	21 22.8	-21 00	75	0.1447	17.6
2347	21 26.7	-22 26	79	0.1196	16.4
2355	21 32.8	+01 10	112	0.1244	17.7
2356	21 33.2	-00 06	89	0.1161	17.1
2377	21 43.3	-10 16	94	0.0808	16.9
2388	21 51.1	+08 00	46	0.0615	16.5
2443	22 23.7	+17 05	117	0.108	16.5
2444	22 24.8	-24 06	76	0.324	17.9
2459	22 33.9	-15 55	33	0.0736	16
2471	22 39.3	+07 00	92	0.1078	17.7
2496	22 48.3	-16 40	104	0.1233	17.2
2559	23 10.5	-13 58	73	0.0796	17
2597	23 22.7	-12 23	43	0.0852	16.6
2616	23 30.7	+05 20	94	0.1832	17.2

Data

Reference: Abell, Corwin, Olowin

Clusters ΔZ = 0. No change for about four decades.

Figure # 30

Figure 30 cmn z 2618-3969.xls					
ACO and S&R Galactic Clusters with common Z's					
Cataloged in 1959 and 1999 = ~ 40 yeqr time lapse					
			max=321	max=.308	max=19.6
ACO #	RAB1950	DEB1950	Count	z	m10
	"h:m:s"	"d:m:s"	ct		mag
2618	23 31.3	+22 44	35	0.0705	15.9
2623	23 32.5	+05 20	142	0.1784	17.2
2625	23 33.8	+20 15	45	0.0609	15.6
2632	23 35.2	-09 30	144	0.186	17.8
2638	23 37.9	-11 59	123	0.0825	17.2
2646	23 38.8	-10 17	135	0.193	17.6
2658	23 42.4	-12 35	143	0.185	17.4
2665	23 48.2	+05 50	34	0.0556	15.8
2686	23 56.8	-21 05	50	0.1124	16.9
2703	00 02.8	+15 49	46	0.1144	17.1
2731	00 07.7	-57 16	39	0.0312	15.1
2744	00 11.8	-30 40	137	0.308	19.5
3207	04 00.8	-27 20	77	0.212	19.2
3338	05 21.3	-48 19	31	0.0446	17.3
3593	14 16.4	-19 15	32	0.1196	16
3665	20 06.0	-53 19	120	0.237	19
3676	20 22.0	-40 31	33	0.0404	18.1
3685	20 28.3	-56 36	30	0.062	18.1
3687	20 29.1	-63 12	46	0.0759	16.3
3736	21 00.0	-43 31	35	0.0487	18
3771	21 26.1	-51 02	42	0.0796	16.1
3783	21 30.8	-42 52	118	0.1955	18.5
3826	21 56.5	-56 24	62	0.0754	14.9
3831	22 00.2	-46 04	81	0.065	19
3849	22 12.7	-51 48	42	0.0678	15.8
3934	22 50.8	-33 59	113	0.224	19.3
3969	23 02.8	-44 25	55	0.0699	17

Appendix 6

Pertinent Equations, Nomenclature and data:

$H = v / d$; Hubble's Law (asserts straight line acceleration)
But it is ~ valid only for describing the d of individual
astronomical objects from their 'v' measurement.
Reference [43] (See figures 22 and 23)

$a = \Delta v / t = (v_1 - v_2)/t$; acceleration, OK only for the same object
$v_1 = c\, [\{(Z_1 + 1)^2 - 1\} / \{(Z_1 + 1)^2 + 1\}] = \underline{\qquad}$Km / sec
$d_1 = v_1 / H\# = \underline{\qquad}$ light years (good estimate to v = .5 c)
$d_2 = \{d_1 + [(v_1)\,(t_2 - t_1)]\} = \underline{\qquad}$ light years
$Z = [(\lambda_2 - \lambda_1)/ \lambda_1] = \underline{\qquad}$ no units
    Wave Length (λ) spectroscope measured in Angstroms
$\Delta Z = [Z_1 - Z_2] = [(\lambda_1 - \lambda_0) / \lambda_0] - [(\lambda_2 - \lambda_0) / \lambda_0]$
H# appears to vary dependent upon direction: (50 to 117, [1])
H # = @ 75 Km / sec / Megaparsec, (1 parsec = 3.261633 ly)
   = @ 75 Km / sec / 3,261,633 ly (or / .003261633 Bly)
H# = 75 Km / sec / Mps = 0.00002299462 Km / s / ly
H # = 75 Km / sec / Mps = 22,994.62 Km / sec / Bly,
                        $\approx$ 23,006.00 Km / sec / Bly

Essential numbers:
$c \sim = 299{,}792$ Km / sec
λ = wave length, in Angstroms, Å, = 10^{-8} cm
Z Red shift ranges from 0 to ∞, at c
4 decades   = 1,262,476,800 sec
1 year   = 31,561,920 sec
40 light years   = 378,735,040,000,000. Km
                 = $3.7873504000 \times 10^{14}$ Km
1 light year = 9.4683760×10^{12} Km
1 Bly   = 9.4683760×10^{21} Km
Relative speeds: (max) $\pm$; Object speed $\pm \sum$ (A.$\pm$B.$\pm$C.$\pm$D.$\pm$E.)
(All are vector / directional sensitive)
A. Earth spin tangential; 1000 mi /hr, = xxx Km/sec
B. Earth revolving tangential; 326725360 mi/yr, = xxx Km/sec

C. Solar sys tangential; 175929.04 lightyears /220,000,000 yrs
    = xxx Km/sec

D. Milky Way Tangential; ???

E. Local Group Tangential; ???

Pertinent Laws of Physics for universe origin

Figure # 1 repeat

LAWS of PHYSICS Pertinent to Origin of the Universe

Apply at all times and places without limit. Cannot be violated.

All mankind's knowledge and understandings are based on Scientific Thinking.

Scientific Thinking Uses logic only with proven facts & verifiable phenomena
Laws of Physics Principles verified many times, but never disproved
Myths Cannot be proven or disproved. (Based on fantasy or wishful thinking)
Belief & Faith ...Accepting without proof. (e.g., BB)
Presumption ...Arrogant acceptance of belief based on assumption or supposition

Proven, and never disproved:

(1)Motion (Isaac Newton, 1610)
$F = ma, = d(mv)/dt$
Force required for acceleration

(2)Energy & Mass (Albert Einstein)
$E = mc^2$
Energy and mass are convertible

(3)Force of Gravity (Isaac Newton)
$F_g = m_1 \times m_2/d^2$
Mutual attraction all objects

(4)Relativity (Albert Einstein, 1905)
$m = m_0 / (1 - m^2/c^2)^{1/2}$

(5)Continuity (Rudolf Clausius, 1850)
1st law of thermodynamics
$M_1 = M_2$, mass & energy into a
system = mass & energy out + mass
& energy remaining in the system.
Succinctly; Matter cannot be created or destroyed. (No BB!)

(6)Schwarzschild radius (1610)
$R_S = 2Gm / c^2$
Radius where escape velocity = $< c$

(7)Escape Velocity
$V_{esc} = (2Gm / r)^{1/2}$,
For a black hole: $*V_{esc} = c$
(*Impossible to achieve)

Nomenclature:
F ... force
E ... energy
m ... mass
v ... velocity (speed)
V ... velocity (vector)
a ... acceleration ($\sim dv/dt$)
c ... velocity of light (light speed)
M ... matter (both mass & energy)
e ... entropy
r ... radius
G ... universal force of gravity
$= 6.67 \times 10^{-11}$ ... Nm^2 / kg^2
N ... Newton unit of force
(Kg meters / sec^2)
R_S ... Schwarzschild radius

Appendix 7

Figure files

Word/Supplement/Model Images

Appendix 8

History-Astronomical Thinking-Summary

Millennium 1 (Until ~ 1000 AD)
Earth/world/universe was believed to be flat
Early myth / thoughts for origin of the universe; a giant and a cow got together and laid an egg. When it hatched, there was the universe! Carl Sagan reported this history in his book "Cosmos". By the end of Millennium 1, some scientific thinking had begun.

Millennium 2 (From ~ 900 AD to ~ 2005 AD)
Earth/world and the universe are known to be dynamic.
Technology advanced. Red shift studies show far away galaxies are separating faster than close ones. That was misinterpreted to explain closer galaxies are accelerating away. By "back tracking" to what was interpreted as the beginning, the BB theory was conceived; Laws Of Physics were ignored and unprovable hypothetical phenomena were 'presumed' for explanations.

Millennium 3 (After ~ 2005 AD)
With this Universe Model explanations are scientific and within the Laws Of Physics; unprovable ideas are rejected for any part of the explanation. Rejected concepts are ideas such as: "dark energy" expansion without force; early-on "faster than light"; and matter (mass and energy) erupting from nothing. The current (2010) "Model of the Universe" is scientifically credible and Laws of Physics compliant interpretations and explanations can now prevail for all astronomical studies.

This documentation includes new analyses, also it is a refinement and clarification of the concepts described in the NUT [1]. It is totally consistent with the laws of physics, and uses no mythical or hypothetical concepts such as 'faster than speed of light' and 'dark energy'. Scientific thinking will hopefully prevail in the future, and then for evermore. (Scientific thinking uses logic with only proven facts, and verifiable phenomena.)

Appendix 9

Red-Shift
Wikipedia Encyclopedia Input:

The following discussion of red shift is, as it is proposed for use in "the Wikipedia free encyclopedia". Their present explanation includes interpretations and explanations about theories that are hypothetical.

Red-shift is a tool which for astronomical purposes provides an accurate way to measure the separation velocity between us and a distant galaxy at the time the light waves left their originating galaxy. **Red-shift** data also proves the farther away galaxies are traveling faster than closer ones thus we know galaxies are separating from each other. Not directly measurable is whether the separation speeds are steady, decreasing, or increasing. As of 2009, demonstrating that the galaxies' velocities of separations are changing with time has not been accomplished. Separations are increasing because velocities are different, but if the rates are constant or decreasing, the universe can be said to be 'growing', but not 'expanding'. There is a very significant difference. To solve this mystery will require taking two or more time separated **red-shift** measurements with adequate red-shift resolution, from the same light originating galaxy, to show if the separation velocity is increasing, steady, or decreasing;

A. If the separation velocities are decreasing, it is because galaxies are continuously transferring some of their linear momentum to angular momentum. This is to be expected because they intermittently and irreversibly cluster together due to relentless gravitational attractions, and subsequently mutually orbit.

Appendix 9 continued

Integration of a galaxy into a cluster likely requires many thousands of years, and occurs intermittently, therefore two **red-shift** velocity measurements of the same galaxy might only show the same **red-shift** and separation velocity, even though over the long haul the velocity may be decreasing. If the two or more red-shift measurements show either no increase in separation velocity or decreasing separation velocity, it will disprove 'expansion' as defined in this article.

B. The 1920's astronomers envisioned the separation velocities as increasing with time only because separations are increasing with distance. They theorized that a hypothetical 'dark energy' force is causing acceleration and continuously pushing the galaxies apart. To prove if the universe expansion theory is real and the separation velocities are, or are not, increasing, two or more time separated **red-shift** measurements of the same galaxy are required to prove increasing **red-shift** from the same galaxy. Such measurements have not been achieved; nevertheless that is what the BB Theory proponents presume, even though that assumption requires a hypothetical dark energy phenomena.

Red-shift studies by astronomers in the 1920's assumed not only galactic separation distances are increasing, but they also presumed the velocities of galaxies are increasing. That led to the additional presumption that galaxies all originated from the same point, time, and source, which was named the Big Bang (BB) [61]. The BB Theory single point origin assumption requires violation of several proven laws of physics as well as postulation of phenomena that cannot be proved or disproved; e.g., dark energy, universal expansion, early hypothetical 'faster than light speed' expansion called inflation [62] , an original infinite energy source, disregard for Shwarzschild radii, etc.

<u>Appendix 9</u> continued

Since the year 2000, Galactic **red-shift** observations have been reconciled with other observations using only proven facts, and verified phenomena which are the Laws of Physics. Scientific logic therefore concludes galaxies are probably decelerating and red-shifts are decreasing with time. All objects would have precipitated from a transient wave mechanism which transformed primordial matter into the mass objects we observe, and all objects continue to decelerate via coalescing and vortexing, which involves transfer of linear momentum to angular momentum [63]. This plausible concept [64] explains primordial matter and how it transforms into energy and then further transforms [65] into elementary particles, neutrons, protons, nuclides, and then into stars, and galaxies. All inter-galactic observations demonstrate the universe is rich with angular momentum [66] and stars and galaxies continue compounding and clustering [66]. The Abell galactic cluster catalogue [67] lists **red-shifts** and directions for 4,076 galaxy clusters with between 29 and 322 galaxy count in each. APOD #2003/08/03 Ref [68] is a photograph of a small sample of untold numbers of rotating galaxies.

Red-shift studies with large modern telescopes are proposed to provide proof the universe is continuing to grow while galaxies and galactic clusters are decelerating, which will undeniably disprove the BB expansion theory. In due time the **red-shift** verified concept that is 'Laws of Physics compliant' will then become the main stream among all Scientific Thinkers (Users of logic with only proven facts and verified phenomena).

Appendix 9 continued

References:

61. "History of the Origin of the Chemical Elements …", 2004, Norman E. Holden, Brookhaven National Laboratory, Upton NY 11973-5000, USA

62. "The Inflationary Universe; Quest for a New Theory of Cosmic Origins" by Alan H Guth. 1997, ISBN 0-201-14942-7, Perseus Books, Cambridge, MA

63. "New Universe Theory with Laws of Physics", ISBN: 1-4184-9430-5 (dj) © 2005 Bobby McGehee Library of Congress control Number: 20055904178; NUT Concept, from primordial matter to the present universe; pages 59 to 63.

64. "New Universe Theory with Laws of Physics", ISBN: 1-4184-9430-5 (dj) © 2005 Bobby McGehee Library of Congress control Number: 20055904178 NUT Stages I, II, & III, pages 63 to 117.

65a. Beijing Electron Positron Collider (BEPC); A modern multi-billion dollar laboratory facility that demonstrates matter conversion from Positrons and Electrons into energy and then into elementary particle mass objects Web site:
<www.ihep.ac.cn/English/E-BEPC/index.htm>

65b. "First three Minutes", by Steven Weinberg; First technical analysis / description of the 'Big Bang Model'; first printing February 1977; published by Bantam / Basic Books, ISBN 0-553-14131-7.

Appendix 10

Model Universe Statistics
(Based on Universe Model with Laws of Physics)

Earth Diameter.. 8,000 miles
    Distance to Sun.....................................92 m miles
    Nearest planet to sun (Mercury)..........45 m miles
Solar System ....1 m miles
    Planets (- Pluto?).................................... 9
Nearest other star (Centauri).......................4.3 ly
    (3 star system, Alpha Beta & Proxima)
Milky Way Galaxy diameter.......................105,000 ly
    Estimated Star count .... 400 billion
    Distance to center .... 28,000 ly
    Distance to edge ....30,000 ly
    Thickness of disk..................................150 ly
Local Group Galactic Cluster members..... ~30
    Nearest Neighbor (Andromeda)...........2.2 m ly
Nearest Clusters .... #'s 2640, 2641, 3061,
and 4067
    Cataloged Count by ACO.....................64 to 184 ct
    Recessional velocity..............................0109 Z = ~3250 km/s
    Distance to alpha (ref) galaxy ....1413341339 Bly
*Universe Diameter (including D wave) ..... ~ 60 Bly
    Origin to Earth (min)............................~ 3.5 Bly
    Deflagration wave thickness................. ~ 2.5 Bly
    Age of Universe.................................... ~ 30 B years
    Distance to Deflagration wave.............. ~ 24 Bly
    Observable objects to (max).................~14 Bly
    Estimated Galaxy Count .... > ~ 1 Trillion
    Primordial Matter, distance .... ~ 26.5 Bly
    Primordial Matter Extent .... Endless !?**

*Model Universe data from Figures #3 & #20
** Three puzzles left for quandary:**
    1. How did primordial originate?
    2. What triggered the initial annihilation?
    3. What is the extent of primordial matter?

Credits: *(Alphabetical)* **All contributors are appreciated.**

Virginia L (nee McGehee) Almquist, BA, MA, English, Rtd.
Contribution: Punctuation and grammar editing.

James C Baker, ME, Rocket Engineering Scientist, Rtd, Boeing, NASA, Huntsville, AL
Contribution: Suggestions for a specific and implicit valid Red-Shift rewrite for Encyclopedia Wikipedia. (Appendix 9)

Harold E Corwin, PhD, Prof Physics & Astronomy, Calif Inst of Technology, Original Member of Abell Corwin Olowin (ACO) Galaxy Cluster catalogers / researchers
Contribution: Suggestions for Galaxy Cluster studies and provided references to Struble & Rood catalog data.
(Appendix 5, Figure 9, 10, 11, & 12)

Robert E Farrell, Dr & Professor Mechanical Engineering, Rtd, Univ of Mass, Lowell, Rtd Penn State, (most recent assignment at Penn State University)
Contribution: Recommendations for representative / realistic Hubble gradient line to be asymptotic to the deflagration wave at .97+ c (Figure 18, 19, & 20)

John E Huchra, PhD, Prof and Astronomy Researcher, Harvard Univ, Mass
Contribution: Informed me about the "Dark Energy' study teams and for communication purposes, supplying the names of participants. Team leader: Adam Reiss (Figure 5, 6, & 7)

Wm G Ryder, Nuclear Engineer, Oklahoma University; Palo Verde Nuclear Plant, AZ
Contribution: Assisted in calculations of Universe Growth to disprove hypothetical accelerating Expansion rates. (see '5 step'; Proof of no 'dark energy' and that Hubble relationship is only a

position description, not a math equation!) (Appendix 2, Figure 21 & 22)

Herbert J Rood, Princeton NJ,
Mitchell F Struble, Lockheed Martin PA
Contribution: Struble & Rood Published "Velocity dispersions for ACO clusters"; The Astrophysical Physical Journal Supplement series 125-35 – 71, 1999. Their cataloged data compared to ACO data for the same galactic clusters (acquired 40 years earlier) provides the basis for verifiable proof the universe is not expanding with acceleration.

Model of the Universe

Addendum

Bobby McGehee explains: This Model of the Universe is based on Logic using proven Facts and the Laws of Physics (which are phenomena verified many times and never disproven). This Model is verified by documented observations and discoveries made by credible astronomers and physicists (see references and credits). This Model uses accepted and proven: Isaac Newton's 1620 Gravity; Albert Einstein's 1909 velocity/mass Relativity; V M Slipher's 1912 Red Shift discovery that the farther away the next galaxy, the faster it is receding; and Paul Dirac's 1927 -1931 discovery of Positrons and Positroniums. Therefore this is not a new theory, although it was originally labeled as such. It is a totally provable concept (Appendix 2).

McGehee, along with many other scientific thinking people never accepted the BB theory because the BB is based on unfounded and unprovable ideas that are not consistent with established Laws of Physics.

About the Author

Career Engineering-Physicist Bobby L McGehee was born in 1927 and reared in Enid Oklahoma. He served in the USMS and USAF, Studied at Oklahoma State University, qualified for three degrees: Engineering-Physics, Physics, and Secondary Education. While taking graduate courses he was employed by the Head Professor of the Physics Department to teach undergraduate physics. During his exciting 32 year career in the aerospace industry he was honored with Engineer-of-the-Month and Engineer-of-the-Quarter awards. He was an Associate Fellow of the American Institute of Aeronautics and Astronautics, and served on the Board. His career was Research and Management for developing, building and operating Aerodynamic and Propulsion Test Facilities, including; Boeing's first Supersonic Wind Tunnel for Propulsion testing, Boeing Wichita's first Jet Engine Test Stand, the Large Anechoic Test Chamber, and many others. After retirement he returned to college for three years to study Astronomy, Chemistry, and Geology. He is currently a member of the Astronomy Society of the Pacific, American Association of Physics Teachers, and American Association for the Advancement of Science. He is an avid student of scientific discoveries, and often attends local and national scientific lectures.

Reasons why the BB concept can not be valid:

1. The universe is far older than possible with the BB theory.

2. The BB concept is inconsistent with at least six pertinent Universal Laws of Physics.

3. Orderly systematic distribution of Galactic Cluster counts reveals universe growth direction.

4. Red Shift measurements increase with distance, not time; therefore there are no reasons to assume acceleration.

5. Universe's total mass including dark matter far exceeds possible BB theory origin processes.

__Conclusion__

"The <u>Known</u> universe includes only known facts"

__Purpose:__

This <u>Model of the Universe</u> is based only on factual knowledge (<u>proven</u> facts and phenomena) and thereby realistically describes the universe.

__Knowledge:__

1. There is one universe.
2. There is one time dimension.
3. There are three spatial dimensions.
4. Matter of the universe includes mass and energy. Mass and energy can be transformed ($\leftrightarrow$).
5. 'Laws of Physics' govern all physical processes   and have been proved many times, but never disproved. Laws of Physics apply at all times and places.
6. Scientific thinking uses deductive logic with only proven facts and verifiable phenomena. (Pertinent "Laws of Physics" Figure 1 & 26).

Hypotheses, fiction, and fantasy are fun, and can aid new theory development. Facts require proof.

END